MISSISSIPPI RIVER WATER QUALITY AND THE CLEAN WATER ACT

Progress, Challenges, and Opportunities

Committee on the Mississippi River and the Clean Water Act

Water Science and Technology Board

Division on Earth and Life Studies

NATIONAL RESEARCH COUNCIL
OF THE NATIONAL ACADEMIES

THE NATIONAL ACADEMIES PRESS
Washington, D.C.
www.nap.edu

THE NATIONAL ACADEMIES PRESS 500 Fifth Street, N.W. Washington, DC 20001

NOTICE: The project that is the subject of this report was approved by the Governing Board of the National Research Council, whose members are drawn from the councils of the National Academy of Sciences, the National Academy of Engineering, and the Institute of Medicine. The members of the committee responsible for the report were chosen for their special competences and with regard for appropriate balance.

Support for this project was provided by the McKnight Foundation under Grant No. 05-083. Any opinions, findings, conclusions, or recommendations expressed in this publication are those of the author(s) and do not necessarily reflect the views of the organization that provided support for the project.

International Standard Book Number-13: 978-0-309-11409-7
International Standard Book Number-10: 0-309-11409-8

Additional copies of this report are available from the National Academies Press, 500 Fifth Street, N.W., Lockbox 285, Washington, DC 20055; (800) 624-6242 or (202) 334-3313 (in the Washington metropolitan area); Internet, http://www.nap.edu.

Printed in the United States of America.

THE NATIONAL ACADEMIES

Advisers to the Nation on Science, Engineering, and Medicine

The **National Academy of Sciences** is a private, nonprofit, self-perpetuating society of distinguished scholars engaged in scientific and engineering research, dedicated to the furtherance of science and technology and to their use for the general welfare. Upon the authority of the charter granted to it by the Congress in 1863, the Academy has a mandate that requires it to advise the federal government on scientific and technical matters. Dr. Ralph J. Cicerone is president of the National Academy of Sciences.

The **National Academy of Engineering** was established in 1964, under the charter of the National Academy of Sciences, as a parallel organization of outstanding engineers. It is autonomous in its administration and in the selection of its members, sharing with the National Academy of Sciences the responsibility for advising the federal government. The National Academy of Engineering also sponsors engineering programs aimed at meeting national needs, encourages education and research, and recognizes the superior achievements of engineers. Dr. Charles M. Vest is president of the National Academy of Engineering.

The **Institute of Medicine** was established in 1970 by the National Academy of Sciences to secure the services of eminent members of appropriate professions in the examination of policy matters pertaining to the health of the public. The Institute acts under the responsibility given to the National Academy of Sciences by its congressional charter to be an adviser to the federal government and, upon its own initiative, to identify issues of medical care, research, and education. Dr. Harvey V. Fineberg is president of the Institute of Medicine.

The **National Research Council** was organized by the National Academy of Sciences in 1916 to associate the broad community of science and technology with the Academy's purposes of furthering knowledge and advising the federal government. Functioning in accordance with general policies determined by the Academy, the Council has become the principal operating agency of both the National Academy of Sciences and the National Academy of Engineering in providing services to the government, the public, and the scientific and engineering communities. The Council is administered jointly by both Academies and the Institute of Medicine. Dr. Ralph J. Cicerone and Dr. Charles M. Vest are chair and vice chair, respectively, of the National Research Council.

www.national-academies.org

COMMITTEE ON THE MISSISSIPPI RIVER AND THE CLEAN WATER ACT

DAVID A. DZOMBAK, *Chair*, Carnegie Mellon University, Pittsburgh, Pennsylvania
H. H. CHENG, University of Minnesota (emeritus), St. Paul
ROBIN K. CRAIG, Florida State University, Tallahassee
OTTO C. DOERING III, Purdue University, West Lafayette, Indiana
WILLIAM V. LUNEBURG, JR., University of Pittsburgh, Pittsburgh, Pennsylvania
G. TRACY MEHAN III, The Cadmus Group, Inc., Arlington, Virginia
JAMES B. PARK, Consultant, Loami, Illinois
NANCY N. RABALAIS, Louisiana Universities Marine Consortium, Chauvin
JERALD L. SCHNOOR, University of Iowa, Iowa City
DAVID M. SOBALLE, U.S. Army Corps of Engineers Research and Development Center, Vicksburg, Mississippi
EDWARD L. THACKSTON, Vanderbilt University (emeritus), Nashville, Tennessee
STANLEY W. TRIMBLE, University of California, Los Angeles
ALAN H. VICORY, Ohio River Valley Water Sanitation Commission, Cincinnati

National Research Council Staff

JEFFREY W. JACOBS, Study Director
ANITA A. HALL, Senior Program Associate

Preface

The Mississippi River has long been one of the great defining natural features of the United States. "Mississippi" is an Ojibwa (Chippewa) Indian word meaning "great river" or "gathering of waters." The first recorded European to see the Mississippi River was Hernando de Soto, who led a Spanish expedition across the river in 1541. In their search for a Northwest Passage, Marquette and Joliet traveled on the river in 1673. Shortly after the Louisiana Purchase, while Lewis and Clark were leading the Corps of Discovery up the Missouri River and to the Pacific Ocean, U.S. Army Lieutenant Zebulon Pike was leading a military reconnaissance expedition up the Mississippi River in the summer of 1805. Later, during the steamboat era of the 1800s, Samuel Clemens traveled the river and began writing his impressions of steamboating and river life under the pen name of Mark Twain.

In addition to the rich history and culture surrounding the Mississippi River, the length of the river and the extent of its basin are exceptional and part of the river's uniqueness. It is one of the world's largest rivers in terms of both length and basin size. The basin encompasses almost half the area of the continental United States and contains many different ecosystems, climate zones, and land uses. Several of the Mississippi's tributaries, such as the Arkansas, Missouri, Ohio, White, and Wisconsin Rivers, are large rivers themselves.

Given the Mississippi River's value as a transportation corridor, the development and maintenance of a navigable river channel has long been a primary focus of commercial navigators and the U.S. government. The U.S. Army Corps of Engineers began its efforts on channel improvements and snag removals in the 1800s, and in the 1930s the Corps constructed the locks and dams on the upper Mississippi River that support the current 9-foot minimum channel depth for navigation on the upper river. Further downstream, the Corps of Engineers has been involved in many other river control and channel maintenance activities, including the construction and maintenance of large Mississippi River levees and the Old River Control Structure at the divergence of the Atchafalaya and Mississippi Rivers.

In contrast to the long-standing efforts to control the Mississippi River for navigation and flood management, concerns about water quality in the Mississippi River are more recent. The Clean Water Act of 1972 and its subsequent amendments have been the driving forces of efforts over the past three decades to monitor, characterize, and take steps to improve water quality in the Mississippi River. The Clean Water Act has resulted in many improvements in Mississippi River water quality. Many point source discharges of liquid and solid pollutants to the river, such as municipal sewage systems and industrial plants, have been brought under control through regulated effluent limits, resulting in marked improvements in water quality. During the 35 years of Clean Water Act implementation, the focus of activity has been on point source discharges through the issuance and monitoring of discharge permits. Diffuse, nonpoint sources such as runoff from urban and agricultural lands have received much less attention. These sources contribute nutrients, sediments, toxic substances, and other materials to the river and have proven more challenging to control than point sources.

The 10 states along the Mississippi River corridor differ in the extent to which they have focused on monitoring and assessing water quality in the Mississippi River compared to other waterbodies within their states. For the most part, their Clean Water Act implementation efforts have focused on streams and rivers contained entirely within state borders. Large interstate rivers such as the Mississippi present special challenges for effective Clean Water Act implementation.

Long-standing and growing concerns of a number of groups about lack of coordination among states in implementing Clean Water Act provisions for protection and improvement of water quality in the Mississippi River prompted the McKnight Foundation of Minneapolis, Minnesota, to request the National Research Council (NRC) to undertake a study of the issue. The Committee on the Mississippi River and the Clean Water Act was appointed in 2005 by the NRC and conducted its deliberations and its report production in response to the Statement of Task in Box 1-1.

The committee examined how effectively the Clean Water Act has been applied in terms of protecting and restoring the water quality of the Mississippi River and how its provisions might be used even more fully. The committee did not undertake an examination of the adequacy of the law itself. All discussions and investigations were conducted in the context of the existing Clean Water Act, with the presumption that it will not be changed substantively in the foreseeable future.

The committee held meetings in 2005 and 2006 in four cities along the Mississippi River: Minneapolis, Dubuque, St. Louis, and Baton Rouge. The committee also convened one meeting at the National Academies offices in Washington, D.C. These meetings included presentations by representatives from universities, federal and state agencies, regional stakeholder groups, and members of the public (Appendix A lists guest speakers invited to the committee's meetings). In addition to oral presentations, written comments from many state agency and interest group representatives and the public were submitted and considered. These presentations and written submittals were of significant value to the committee and made clear that the water quality of the Mississippi River and the northern Gulf of Mexico is a scientific and public policy topic of great regional and national importance.

I thank the members of the committee for their uniform commitment to the endeavor, their good cheer, and their diligent efforts. The committee brought considerable range and depth of experience and expertise to the task. Our interactions were rich and produced insights and recommendations that we hope are valuable for Mississippi River water quality planning. It was a privilege to work with this outstanding group.

I also thank the NRC staff members for their dedication and careful work over the course of the study. Jeff Jacobs, senior staff officer with the Water Science and Technology Board (WSTB), helped keep the committee on task and on schedule. Jeff and I worked collaboratively to organize and guide the committee writing assignments, to compile and edit all written contributions for a coherent consensus report, and to ensure that the views and comments of all committee members were considered in developing the report. Jeff's professional insights and his keen editing skills were most helpful and much appreciated. The committee also was ably assisted by Anita Hall, WSTB senior program associate, who handled logistics for our meetings and various aspects of report draft production and dissemination.

The committee is grateful to our sponsor, the McKnight Foundation, for financial and intellectual support of the project. We extend special thanks to Gretchen Bonfert, environment program director at the foundation, and to her colleague Ron Kroese. Gretchen and Ron were very helpful in suggesting experts and knowledgeable advocates to visit with our committee, and they carefully followed committee activities by attending public sessions of all committee meetings. The McKnight Foundation has focused on water quality in the Mississippi River and the Gulf of Mexico as a funding priority since 1992. Today, McKnight's Water Quality Collaborative, a group of many different organizations along the 10-state river corridor, is working to build coalitions to help improve Mississippi River water quality. The McKnight Foundation is to be commended for its vision and commitment in supporting a National Academies review of this important, complex, and sometimes controversial topic.

This report was reviewed in draft form by individuals chosen for the breadth of their perspectives and technical expertise in accordance with procedures approved by the NRC's Report Review Committee. The purpose of this independent review was to provide candid and critical comments to assist the institution in making its published report as sound as possible and to ensure that the report meets institutional standards for objectivity, evidence, and responsiveness to the study charge. Reviewer comments and the draft manuscript remain confidential to protect the integrity of the deliberative process. We thank the following individuals for their review of this report: Clifton J. Aichinger, Ramsey-Washington Metro Watershed District; William L. Andreen, University of Alabama; Paul L. Freedman, Limno-Tech, Inc.; Jerome B. Gilbert, consultant; Lynn R. Goldman, Johns Hopkins University; Robert H. Meade, consultant; Patricia E. Norris, Michigan State University; Leonard A. Shabman, Resources for the Future; Richard E. Sparks, National Great Rivers Research and Education Center; Robert R. M. Verchick, Loyola University, New Orleans; and Paul D. Zugger, Public Sector Consultants.

Although the reviewers listed above provided constructive comments and suggestions, they were not asked to endorse the report's conclusions and recommendations nor did they see the final draft of the report before its release. The review of this report was overseen by Dr. Frank H. Stillinger, Princeton University, and Dr. Patrick L. Brezonik, University of Minnesota. They were responsible for ensuring that an independent examination of this report was conducted in accordance with institutional procedures and that all review comments were carefully considered. Responsibility for this report's final contents rests entirely with the authoring committee and the institution.

The Mississippi River is a natural and economic resource of inestimable value to the nation. Its water quality affects people and ecosystems and is important to the future of the basin. There are many large-scale and complex challenges associated with Mississippi River water quality protection and restoration. Our committee has worked to consider how these challenges can be addressed within the provisions of the Clean Water Act. We hope that our efforts provide useful advice in meeting the challenges surrounding effective implementation of the Clean Water Act and in enhancing the multiple uses of the Mississippi River for future generations.

David A. Dzombak, *Chair*

Contents

Summary

Flowing approximately 2,300 miles from Lake Itasca to the Gulf of Mexico, the Mississippi River represents a resource of tremendous economic, environmental, and historical value to the nation. The Mississippi River drains the vast area between the Appalachian and the Rocky Mountains, making it the world's third-largest river basin, behind the Amazon and the Congo River basins. The river supports numerous economic and recreational activities including boating, commercial and recreational fishing, tourism, hiking, and hunting. Mississippi River water quality is of paramount importance for the sustainability of the many uses of the river and the ecosystems dependent on it. Numerous cities and millions of inhabitants along the river use the Mississippi as a source of drinking water. Water quality is also important for many recreational and commercial activities. The river's ecosystems and its avian and fish species rely on good water quality for their existence. These ecosystems and the species they support are highly valued and are especially important to communities and economies along the river and along the Louisiana Gulf Coast.

There are many differences between the upstream and downstream portions of the mainstem Mississippi River. Much of the upper Mississippi River is a river-floodplain ecosystem that contains pools, braided channels, islands, extensive bottomland forests, floodplain marshes, and occasional sand prairie. The upper river is home to the Upper Mississippi River National Wildlife and Fish Refuge, which covers 240,000 acres and extends 261 miles along the river valley from Wabasha, Minnesota, to Rock Island, Illinois. Further downstream, many large flood protection levees line the lower river and have severed natural connections between the river channel and its floodplain. There are fewer backwater areas and islands than along the upper river and fewer opportunities for river-related recreation. Moreover, the lower Mississippi River's larger flows and dangerous currents and eddies inhibit river-based recreation and impede water quality monitoring. These upstream-downstream differences affect the nature of water quality problems and the extent of water quality monitoring along the length of the river.

Mississippi River water quality is affected by land use practices, urbanization, and industrial activities across its large drainage basin. Many of these activities, including those that take place hundreds of miles away from the main river channel (or mainstem), can degrade Mississippi River water quality. The establishment of cities and commercial activities along the river has contributed to degraded water quality through increasing pollutant discharges from cities and industry. Congress first enacted the Federal Water Pollution Control Act (FWPCA) in 1948. Congress amended the FWPCA repeatedly from 1956 on; however, substantial amendments in 1972 created the contemporary structure of the act, which acquired the name Clean Water Act in 1977 amendments. An overarching objective of the Clean Water Act is to restore and maintain the chemical, physical, and biological integrity of the nation's waters.

The Clean Water Act has achieved successes in reducing point source pollution, or pollution discharged from a discrete conveyance or pipe (e.g., industrial discharge or a wastewater treatment plant), but nonpoint pollution, which originates from diffuse sources such as urban areas and agricultural fields, has proven more difficult to manage. Despite improvements since passage of the Clean Water Act, the Mississippi River today experiences a variety of water quality problems. Many of these problems emanate from nonpoint pollutant sources. Although the Clean Water Act can be used to address nonpoint source pollution problems, its provisions for doing so have less regulatory authority than its provisions for addressing point source pollution.

This report focuses on water quality problems in the Mississippi River and the ability of the Clean Water Act to address them. Data needs and system monitoring, water quality indicators and standards, and policies and implementation are addressed (the full statement of task to this committee is contained in Chapter 1). The geographic focus of this report is the 10-state mainstem Mississippi River corridor and areas of the Gulf of Mexico affected by Mississippi River discharge. Water quality in the Mississippi River and the northern Gulf of Mexico, however, is affected by activities from across the entire river basin. Comprehensive Mississippi River water quality management programs therefore must consider the sources of pollutant discharges in all tributary streams, as well as along the river's mainstem. This report therefore also discusses landforms, land use changes, and land and water management practices across the Mississippi River basin that affect mainstem water quality.

The committee was not specifically charged to consider possible statutory changes to the Clean Water Act. The committee discussed this topic and chose to conduct its investigations and present its findings and recommendations entirely within the framework of the existing Clean Water Act.

FINDINGS

Mississippi River Water Quality Problems

Numerous human activities across the Mississippi River basin affect the water quality of the mainstem Mississippi River and the northern Gulf of Mexico. These activities include discharges from industries, urbanization, timber harvesting, construction projects, agriculture, and landscaping practices. Along the mainstem Mississippi, major hydrologic modifications implemented over the past 150 years also affect water quality. These modifications include river channelization, locks and dams (and associated navigation pools) of the upper Mississippi River navigation system, many large levees along the lower river, and losses of large areas of natural wetlands.

These activities and modifications contribute to many water quality problems along the river's mainstem that vary and are of different magnitude in different parts of the river. These problems can be divided into three broad categories: (1) contaminants with increasing inputs along the river that accumulate and increase in concentration downriver from their sources (e.g., nutrients and some fertilizers and pesticides); (2) legacy contaminants stored in the riverine system, including contaminants adsorbed onto sediment and stored in fish tissue (e.g., polychlorinated biphenyls [PCBs]; dichlorodiphenyltrichloroethane [DDT]); and (3) "intermittent" water constituents that may or may not be considered contaminants, depending on where they are found in the system, at what levels they exist, and whether they are transporting adsorbed materials that are contaminants. The most prominent component in the latter category is sediment. In some portions of the river system, sediment is overly abundant and can be considered a contaminant. In other places it is considered a natural resource in deficient supply.

Differences in inputs of pollutants in different parts of the river basin contribute to varying water quality problems along the length of the river. For example, downstream sediment loads are greatly affected by sediment inputs from, and retention in, the river's many tributary streams. Nutrients enter the Mississippi River at many points along its course, primarily from nonpoint sources in agricultural areas in the upper Mississippi River basin that are not subject to Clean Water Act permit programs. Nitrogen and phosphorus are nutrients of special concern. These nutrients ultimately are discharged into the Gulf of Mexico, where nitrogen causes large-scale problems in the form of hypoxia and other coastal ecosystem disturbances, including impairment of Gulf fish populations. In other portions of the river system, primarily in the upper river, excessive loadings of phosphorus constitute a problem (e.g., in Lake Pepin in southern Minnesota).

Sediment problems are more complex. For example, in the upper Mississippi River, high rates of sediment input and deposition are key concerns. Sediment loads in the upper river today are greater than they were in the mid- to late eighteenth century, when the basin was being settled by European immigrants. The system of locks and dams and navigation pools put in place on the upper river in the early twentieth century affects sediment transport and deposition significantly. In the lower Mississippi River below Alton, Illinois, deprivation of sediments—due in large part to the trapping of large amounts of sediment behind a series of dams and reservoirs on the Missouri River—is a problem. Sediment deprivation is, for example, a key contributor to losses of coastal wetland systems in southern Louisiana. This problem is enhanced to some degree by extensive levee structures along the lower part of the river that do not allow sediments to spread into and across floodplains and wetlands adjacent to the river and its tributaries.

Identifying the most important water quality problems in the mainstem Mississippi River depends on the scale examined. At the local level, for instance, problems with toxic substances and bacteria may be of primary concern to citizens and regulators. **However, at the scale of the entire Mississippi River, including its effects that extend into the northern Gulf of Mexico, nutrients and sediment are the two primary water quality problems. Nutrients are causing significant water quality problems within the Mississippi River itself and in the northern Gulf of Mexico. With regard to sediment, many areas of the upper Mississippi River main channel and its backwaters are experiencing excess suspended sediment loads and deposition, while limited sediment replenishment is a crucial problem along the lower Mississippi River and into the northern Gulf of Mexico.**

Water Quality Monitoring and Assessment

The Mississippi River serves as a border between states along much of its course from Lake Itasca to the Gulf of Mexico. Some states along the river view Mississippi River water quality as primarily a federal responsibility—especially states in the lower stretch of the river. Many of the 10 states along the river thus allocate only small amounts of funds for water quality monitoring and related activities. Moreover, there is very limited coordination among the Mississippi River states on water quality monitoring activities. The Clean Water Act is relatively clear in delineating responsibilities for state-specific water quality monitoring and assessment; it is less clear in addressing issues of coordinated interstate river monitoring and assessment to ensure that water quality data are collected and analyzed in a consistent fashion. **As a result of limited interstate coordination, the Mississippi River is an "orphan" from a water quality monitoring and assessment perspective.**

The orphan-like nature of the Mississippi River entails several unique water quality monitoring and management challenges. One problem stems from the fact that individual states generally are responsible for monitoring the stretch of the Mississippi River that flows through or abuts them. The Mississippi River flows within only two states—Minnesota and Louisiana—of the ten states along its corridor. For the other eight states, the river forms a boundary between them. Although there are some important federally sponsored efforts in monitoring Mississippi River water quality—such as those conducted by the U.S. Army Corps of Engineers and the U.S. Geological Survey, especially on the upper river—there is no single water quality monitoring program or central water quality database for the entire length of the Mississippi. Thus, there are limited amounts of water quality and related biological and ecological data for the full length of the Mississippi River, especially the lower river. This limited amount of data inhibits evaluations of water quality problems along the river and into the Gulf of Mexico, which in turn inhibits efforts to develop, assess, and adjust water quality restoration activities. Moreover, the limited attention devoted to monitoring the river's water quality is not commensurate with the Mississippi River's exceptional socioeconomic, cultural, ecological, and historical value. **The lack of a centralized Mississippi River water quality information system and data gathering program hinders effective implementation of the Clean Water Act and acts as a barrier to maintaining and improving water quality along the Mississippi River and into the northern Gulf of Mexico.**

Effectiveness of the Clean Water Act

The Clean Water Act (CWA) is the cornerstone of surface water quality protection in the United States. It employs a variety of regulatory and nonregulatory tools designed to reduce direct pollutant discharges into waterways, finance municipal wastewater treatment facilities, protect wetlands, and manage polluted runoff. Congress designed the 1972 act "to restore and maintain the chemical, physical, and biological integrity of the Nation's waters." The act also called for zero discharges of pollutants into navigable waters by 1985 and "fishable and swimmable" waters by mid-1983. The U.S. Environmental Protection Agency (EPA) and the states are primarily and jointly responsible for implementing the act. The U.S. Army Corps of Engineers also plays a role in Clean Water Act implementation, because it shares responsibility with the EPA in the act's Section 404 wetlands permitting program.

The Clean Water Act aims to achieve water quality improvements by requiring categorical technology-based standards for point source dischargers. The Clean Water Act has been effective in addressing many point source pollution problems, such as discharges from industrial sources and publicly owned sewer systems and treatment works. Further improvements in control of point sources of pollution—notably in connection with urban stormwater and combined sewer overflows—are possible. Such changes, however, are likely to have limited effects on mainstem and northern Gulf of Mexico water quality because only approximately 10 percent of Mississippi River nitrogen loading is from point sources.

For waterbodies that remain impaired after the application of technology-based and water quality-based controls of point source discharges, the Clean Water Act requires application of water quality standards and Total Maximum Daily Loads (TMDLs). The TMDL represents both a planning process to implement standards and a numerical quantity for a pollutant load to receiving waters that will not result in violation of state water quality standards within an adequate margin of safety. The Clean Water Act requires states or the Environmental Protection Agency to develop TMDLs for waterbodies that do not meet water quality standards. **The Clean Water Act has been effective in addressing point sources of water pollutants. Notably, however, the Clean Water Act addresses nonpoint source pollution only in a limited, indirect manner. This is a crucial difference given the significance of nonpoint source water pollution throughout the nation and its special importance to Mississippi River and northern Gulf of Mexico water quality.**

The Total Maximum Daily Load framework is a key aspect of the Clean Water Act and is designed, in part, to address nonpoint source pollutants and to protect and restore water quality. The TMDL concept and its implementation have been used to address both point and nonpoint source inputs to many waterbodies in the United States. The TMDL framework is more easily implemented in smaller watersheds within individual states. Larger rivers and rivers with watersheds that encompass multiple states pose significant implementation challenges for the TMDL framework, particularly with respect to nonpoint source pollution. **For TMDLs and water quality standards to be employed effectively to manage water quality in interstate rivers such as the Mississippi, it is essential that the effects of interstate pollutant loadings be considered fully in developing the TMDL.**

A lack of coordination among federal- and state-level efforts, limited federal oversight of CWA implementation, and failure of some states to include the Mississippi River within their state water quality monitoring programs all contribute to the inability of the EPA and the states to address adequately water quality degradation in the Mississippi River and into the northern Gulf of Mexico. The Clean Water Act requires the EPA to establish water quality criteria; oversee and approve state water quality standards and TMDLs; take over the setting of water quality standards and the TMDL process when state efforts are inadequate; and safeguard water quality interests of downstream and cross-stream states. **The Clean Water Act assigns most interstate water quality coordination authority to the EPA. The Clean Water Act also encourages the EPA to stimulate and support interstate cooperation to address larger-scale water quality problems. The act provides the EPA with multiple authorities that would allow it to assume a stronger leadership role in addressing Mississippi River and northern Gulf of Mexico water quality.**

Despite the authority granted to the EPA in the Clean Water Act, one of the nation's key, large-scale water quality problems—the hypoxic zone in the northern Gulf of Mexico—continues to persist. The Gulf hypoxic zone is a large area that clearly is not meeting the CWA goal of fishable and swimmable waters. **The EPA has failed to use its mandatory and discretionary authorities under the Clean Water Act to provide adequate interstate coordination and oversight of state water quality activities along the Mississippi River that could help promote and ensure progress toward the act's fishable and swimmable and related goals.**

Programs and policies designed to achieve improvements in water quality for the Mississippi River and the northern Gulf of Mexico are affected by the following factors:

1. Resolution of many Mississippi River water quality issues is constrained by pre-CWA structural alterations to the river—for example, locks, dams, and levees, and the losses of wetlands—that the Clean Water Act cannot undo;
2. The Clean Water Act contains no authorities that directly regulate nonpoint sources of pollutants;
3. The Clean Water Act specifically exempts agricultural stormwater discharges and return flows from irrigated agriculture from being regulated as point source discharges and does not address agricultural nonpoint source pollution except as it leaves all nonpoint source pollution management to the states;
4. The interstate nature of the Mississippi River poses complications in coordinating water quality standards and monitoring programs among ten states and four EPA regions;
5. Large rivers such as the Mississippi are physically difficult to monitor, evaluate, and characterize; and
6. Pollutant loadings from ten states impact the Mississippi River and extend into the northern Gulf of Mexico.

Many structural and physical changes to the Mississippi River predate passage of the Clean Water Act. Moreover, Congress did not design the Clean Water Act to address every process that affects Mississippi River water quality. The Clean Water Act has been effective in reducing many pollutant discharges from point sources, but other processes such as levee construction, urbanization, and forestry activities affect Mississippi River quality and are not subject to the regulatory provisions of the Clean Water Act. **The Clean Water Act cannot be used as the sole legal vehicle to achieve all water quality objectives along the Mississippi River and into the northern Gulf of Mexico. Nevertheless, the Clean Water Act provides a legal framework that, if comprehensively implemented and rigorously enforced, can effectively address many aspects of intrastate and interstate water pollution, although the emphasis to date has been predominantly on the former.**

Nonpoint Source Pollution and Agriculture

Since agriculture contributes the major portion of nutrients and sediments delivered to the Mississippi River, reductions in pollutant loadings, especially nutrients, from the agricultural sector are crucial to improving Mississippi River water quality. Not all agricultural producers across the river basin contribute equal amounts of nutrients and sediments in runoff. Water quality protection programs thus need not be implemented in every watershed and on every farm to realize substantial water quality improvements further downstream. The careful targeting of programs to areas of higher pollutant loadings could enhance the effectiveness of conservation programs designed to reduce nutrient and sediment runoff.

The U.S. Department of Agriculture (USDA) administers a number of incentive-based programs designed to implement best management practices (BMPs) and/or reduce levels of nutrient and sediment inputs and runoff. USDA programs to reduce environmental impacts of agriculture include the Conservation Reserve Program (CRP), the Environmental Quality Incentive Program (EQIP), and the Conservation Security Program (CSP). These programs aim to balance incentives for crop production with incentives for land and water conservation. Participation is voluntary, but there are financial incentives for implementing BMPs.

A key issue in Midwest agriculture today is the potential increase in crop land and production dedicated to biofuels. Recent interest in biofuels production is encouraging producers to extend and intensify crop production in much of the upper Mississippi River basin. Much of this expanded production is in corn, which entails large applications of nutrient fertilizers. As a result, sediment and nutrient runoff from agricultural land in the upper basin is likely to increase. Although increases in grain production for biofuels, particularly on marginal agricultural lands that contribute high nutrient loads, may have substantial consequences for Mississippi River and northern Gulf of Mexico water quality, these potential impacts have not been fully evaluated.

RECOMMENDATIONS

Agriculture and Mississippi River Water Quality

Effective management of nutrient and sediment inputs and other water quality impacts from agricultural sources will require site-specific, targeted approaches involving best management practices. Existing USDA programs provide vehicles for implementing nonpoint source controls in agriculture, but they will require closer coordination with the EPA and state water quality agencies to realize their full potential for improving water quality. The EPA could assist the USDA to help improve the targeting of funds expended in the CRP, EQIP, and CSP. The national financial investment and scope of these USDA programs is large. A focus on these programs is important because the Clean Water Act does not authorize regulation of nonpoint sources of pollutants such as agricultural lands. Recent developments in the prospects for increased biofuels production, and the increased nutrient and sediment pollutant loads this would entail, provide an even stronger rationale to expedite targeted applications of USDA conservation programs and enhanced EPA-USDA coordination.

Targeting USDA conservation programs to areas of higher nutrient and sediment loadings can lead to BMPs for control of runoff containing sediment and nutrients being implemented on lands that are the primary sources of nonpoint pollutants. This provides an opportunity to strengthen EPA-USDA interagency collaboration: the EPA, for example, can assist USDA in identifying lands that should receive priority and can cooperate with USDA and producers in monitoring changes in water quality and making subsequent adjustments and improvements in nutrient management programs. The U.S. Geological Survey (USGS) also could play an important role in this collaboration by sharing its considerable expertise and data related to water quality monitoring.

It is imperative that these USDA conservation programs be aggressively targeted to help achieve water quality improvements in the Mississippi River and its tributaries. Programs aimed at reducing nutrient and sediment inputs should include efforts at targeting areas of higher nutrient and sediment deliveries to surface water. The EPA and the USDA should strengthen their cooperative activities designed to reduce impacts from agriculture on the water quality of the Mississippi River and the northern Gulf of Mexico.

State-Level Leadership

The 10 mainstem Mississippi River states have different priorities regarding the river and devote different levels of resources to water quality data collection. Broadly speaking, there is a distinction between priorities and approaches of the upper river states compared to the lower river states. One example of these differences is that the upper river states participate in a governor-supported interstate body—the Upper Mississippi River Basin Association (UMRBA). The five upper river state governors established the UMRBA in 1981 to help coordinate river-related programs and policies and to work with federal agencies with river responsibilities. The UMRBA has sponsored discussions and studies on many water quality issues. At a strategic level, the UMRBA represents an interstate commitment to cooperation on river management issues. There is no equivalent organization for the lower river states. The Lower Mississippi River Conservation Committee (LMRCC) is a multistate organization established to discuss issues of river biology and restoration, but it does not have gubernatorial appointees or employ full-time staff like the UMRBA.

Effective water quality protection and restoration requires that the Mississippi River be managed as an integrated system. Working together, the 10 Mississippi River states will achieve far more, with greater efficiencies, than each state working alone. Mississippi River states will have to be more proactive and cooperative in their water quality programs for the Mississippi River if marked improvements in water quality are to be realized. A mechanism for the lower river states to promote this coordination could take different forms, such as a forum for information exchange or an organization with a more formal status. **Better interstate cooperation on lower Mississippi River water quality issues is necessary to achieve water quality improvements. The lower Mississippi River states should strive to create a cooperative mechanism, similar in organization to the UMRBA, in order to promote better interstate collaboration on lower Mississippi River water quality issues.**

EPA Leadership

Several federal agencies maintain programs related to water quality monitoring across the Mississippi River watershed and into the northern Gulf of Mexico. For example, the National Oceanic and Atmospheric Administration (NOAA) collects water quality data for the Gulf of Mexico, the U.S. Army Corps of Engineers oversees the federal-state Environmental Management Program for the upper Mississippi River, and the USGS has collected water quality data for many years at select Mississippi River stations under different monitoring programs. All of these programs have merit, but there is no single federal program for water quality monitoring and data collection for the river as a whole. The past and current approach to water quality management in the Mississippi River is fragmented, with different agencies conducting their own monitoring programs and having different goals. This does not lend itself to a coherent program designed to monitor and consider the Mississippi River as a whole. The Mississippi River, with its extensive interstate commerce, its ecosystems that cross state boundaries, and its effects that extend into the northern Gulf of Mexico, clearly is a river of federal interest. There are compelling reasons for the federal government to promote the monitoring and evaluation of this river system as a single entity.

Better coordination and a greater degree of centralization of water quality monitoring and data collection along the Mississippi River are essential to ensure that similar parameters are being measured consistently along the entire length of the river; that similar methods, units, and timing of measurements are being used along the entire river; and that the placement and operations of monitoring stations are coordinated. There is an adequate scientific basis to undertake an expanded monitoring program for the Mississippi River. Better coordination is fundamental to streamlining federal expenditures and efforts for water quality monitoring along the river and, ultimately, to achieving water quality improvements in the Mississippi River and the northern Gulf of Mexico. This will help ensure an integrated program that enables consistent, science-based decisions about important water quality monitoring issues.

There is a clear need for federal leadership in system-wide monitoring of the Mississippi River. The EPA should take the lead in establishing a water quality data sharing system for the length of the Mississippi River. The EPA should place priority on coordinating with the Mississippi River states to ensure the collection of data necessary to develop water quality standards for nutrients in the Mississippi River and the northern Gulf of Mexico. The EPA should draw on the considerable expertise and data held by the U.S. Army Corps of Engineers, the USGS, and NOAA.

The EPA should act aggressively to ensure improved cooperation regarding water quality standards, nonpoint source management and control, and related programs under the Clean Water Act. This more aggressive role for EPA is crucial to maintaining and improving Mississippi River and northern Gulf of Mexico water quality and should occur at several levels. **The EPA administrator should ensure coordination among the four EPA regions along the Mississippi River corridor so that the regional offices act consistently with regard to water quality issues along the Mississippi River and in the northern Gulf of Mexico.**

Regarding cooperation and communication among the Mississippi River states, **the EPA should encourage and support the efforts of all 10**

Mississippi River states to effect regional coordination on water quality monitoring and planning and should facilitate stronger integration of state-level programs. The EPA has an opportunity to broker better interstate collaboration and thereby improve delivery of Clean Water Act-related programs, such as permitting, monitoring and assessment, and water quality standards development. The EPA should provide a commensurate level of resources to help realize this better coordination. One option for encouraging better upstream-downstream coordination would be through a periodic forum for state and regional water quality professionals and others to identify and act upon appropriate Clean Water Act-related concerns.

There are currently neither federal nor state water quality standards for nutrients for most of the Mississippi River, although standards for nutrients are under development in several states. Numerical federal water quality criteria and state water quality standards for nutrients are essential precursors to reducing nutrient inputs to the river and achieving water quality objectives along the Mississippi River and in the northern Gulf of Mexico. A TMDL could be set for the Mississippi River and the northern Gulf of Mexico. This would entail the adoption by EPA of a numerical nutrient goal (criteria) for the terminus of the Mississippi River and the northern Gulf of Mexico. An amount of aggregate nutrient reduction—across the entire watershed—necessary to achieve that goal then could be calculated. Each state in the Mississippi River watershed then could be assigned its equitable share of reduction. The assigned maximum load for each state then could be translated into numerical water quality criteria applicable to each state's waters.

Regarding cooperation with the Mississippi River states on water quality standards and criteria, **the EPA should develop water quality criteria for nutrients in the Mississippi River and the northern Gulf of Mexico. Further, the EPA should ensure that states establish water quality standards (designated uses and water quality criteria) and TMDLs such that they protect water quality in the Mississippi River and the northern Gulf of Mexico from excessive nutrient pollution. In addition, through a process similar to that applied to the Chesapeake Bay, the EPA should develop a federal TMDL, or its functional equivalent, for the Mississippi River and the northern Gulf of Mexico.**

The actions recommended in this report will not be easy to implement. They will entail a greater degree of collaboration and compromise among interest groups, states, and agencies than in the past. They are, however, necessary if the goals of the Clean Water Act are to be realized and the Mississippi River provided a level of protection and restoration commensurate with its integral commercial, recreational, ecological, and other values.

1

Introduction

Flowing a distance of roughly 2,300 miles through the heart of the continental United States, the Mississippi River is a resource of great economic value, environmental importance, and cultural and historical significance. The Mississippi River also is notable for the size of its drainage basin area: extending over much of the vast expanse between the Appalachian and the Rocky Mountains, it is the world's third-largest river basin. With an area of more than 1.84 million square miles, it covers approximately 40 percent of the conterminous United States. The basin extends over all or part of 31 U.S. states and two Canadian provinces.

The Mississippi River has long been important for commercial transportation and navigation. Today, hundreds of millions of tons of commodities are shipped annually on the river, and the Greater Port of New Orleans and Baton Rouge handles more grain tonnage than any other port in the world (Port of New Orleans, 2006). The river and its floodplain ecosystem also provide numerous environmental goods and services, such as the Upper Mississippi River National Wildlife and Fish Refuge, the longest river refuge in the continental United States. The river's extensive and multiple values prompted the U.S. Congress to pass the Upper Mississippi River Management Act of 1986, which designates the upper river "as a nationally significant ecosystem and a nationally significant commercial navigation system" (P.L. 99-662). The Mississippi River also is used as a source of drinking water for millions of people in cities along the river. Ensuring adequate water quality in the Mississippi River clearly is a national concern. Despite the importance of the river and its water quality, however, more effective Mississippi River water quality monitoring and management are confounded by administrative, historical, environmental, and other factors.

Water quality programs for the Mississippi River are administered in accordance with the federal Clean Water Act (CWA). At the federal level, Clean Water Act implementation is the responsibility of the U.S. Environmental Protection Agency (EPA). States that border the river also have significant CWA-related management responsibilities, for many of which they have been delegated authority by the EPA. The fact that the Mississippi River flows through or borders on 10 different states (Figure 1-1) is

FIGURE 1-1 Mainstem Mississippi River. © International Mapping Associates.

of no small consequence in implementing provisions of the Clean Water Act. These 10 states have different economies and land uses, different geographical locations along the river, different social and economic priorities, and different levels of fiscal resources, all of which affect their Mississippi River water quality monitoring programs. In addition, the physical, chemical, and biological characteristics of the river vary tremendously along its length, leading to different water quality issues and monitoring challenges in each state. Water quality issues in and along the river's mainstem are further affected by land and water systems and practices in other states in the Mississippi River basin, for example, in upstream areas of large tributaries such as the Arkansas, Missouri, and Ohio Rivers.

Mississippi River water quality monitoring and remediation efforts are also affected by historical and ongoing environmental changes along the river and through its river basin. Many of these changes have affected large areas and date back many decades or longer. On the upper river, construction of the series of navigation dams and pools in the 1930s as part of the 9-foot upper Mississippi River channel project initiated regional-scale hydroecological changes that continue today. Farther downstream, much of the lower river is separated from its natural floodplain by levees and other protective structures, some of which have been in place for 100 years. Losses of natural wetlands, which have been significant in all 10 states along the Mississippi River, date to nineteenth century efforts to drain these areas for agricultural and other human uses and are widespread. In addition to these historical changes to river hydrology and floodplain land uses, toxic substances (e.g., polychlorinated biphenyls [PCBs], dichlorodiphenyltrichloroethane [DDT]), some of which persist in river sediments to this day, were discharged into the river prior to implementation of the Clean Water Act. There are also some continuing inputs of PCBs, DDT, and other toxic substances from stormwater sewers, runoff from urban streets and agricultural lands, and other sources due to the presence of residual contamination.

Because there have been no periodic, large-scale assessments of the long-term status of legacy contaminants buried in the river's sediments and carried in the tissues of the river's biota, assessing the impacts of these legacy factors on the attainment of designated uses along the river has proven difficult. Many of these significant environmental changes to the river predate passage of the Clean Water Act, which was enacted in its present form in 1972. Remedial efforts aimed at improving water quality are often affected by these long-standing changes and thus are not readily addressed within the Clean Water Act framework.

Many different federal, state, and nongovernmental groups have noted and studied these water quality issues and the challenges they pose to implementing the federal Clean Water Act and protecting Mississippi River water quality. Questions about the effectiveness of the Clean Water Act and its implementation along the 10-state Mississippi River corridor prompted the McKnight Foundation of Minneapolis to request the National Research Council (NRC) to conduct a study of CWA implementation along the Mississippi River. The McKnight Foundation and its Environment Program have long-standing interests in Mississippi River water quality. McKnight has also focused on building coalitions among varied groups along the river to help protect and restore water quality. With support from McKnight, the NRC's Water Science and Technology Board (WSTB) convened the Committee on the Mississippi River and the Clean Water Act to conduct

BOX 1-1
Committee on the Mississippi River and the Clean Water Act Statement of Task

The study will review the experience of Clean Water Act implementation along the 10-state Mississippi River corridor. Part of the evaluation will include the review of the following reports: several GAO (formerly the General Accounting Office; now the Government Accountability Office) documents published in the late 1990s and early 2000s on Clean Water Act implementation, a January 2004 report from the Upper Mississippi River Basin Association on Mississippi River water quality, a November 2004 report from the Headwaters Group regarding Clean Water Act implementation on the Mississippi River, and a February 2003 petition to the U.S. EPA from the Sierra Club and EPA's June 2004 response. These documents focus on issues of data collection, water quality, and programmatic and management challenges in implementing Clean Water Act regulations along the Mississippi River and will serve as a point of departure for the committee's deliberations.

Many concerns regarding Clean Water Act implementation and enforcement along the Mississippi River relate to the adequacy of water quality data, challenges involved in establishing system-wide indicators and quality standards, compliance with standards and regulations, and interstate and interagency coordination and cooperation. The committee's statement of task will accordingly be divided into four broad areas: Mississippi River Corridor Water Quality Problems, Data Needs and System Monitoring, Water Quality Indicators and Standards, and Policies and Implementation.

1. Mississippi River Corridor Water Quality Problems: Identify generally the key water quality problems through the 10-state Mississippi River system. The depiction of these problems should not only reflect varying conditions along the river corridor, but also discuss implications of Mississippi River water quality for conditions in the Gulf of Mexico.

a comprehensive study of the act's administration along the river (Box 1-1 contains the committee's statement of task).

MISSISSIPPI RIVER WATER QUALITY ISSUES

Mississippi River hydrology and ecology have changed fundamentally over the past 200 years. These changes have been driven by numerous activities including the construction of locks, dams, and levees, drainage of wetlands, agriculture, urbanization, and timber harvesting—all of which have affected Mississippi River water quality. In addition, many years of

2. Data Needs and System Monitoring:

• Identify and discuss key water quality data needs for the 10-state Mississippi River corridor with regard to Clean Water Act reporting requirements. What are the main barriers to collecting and reporting these data? How could water quality data collection activities and programs for the Mississippi River corridor be revised to promote progress toward Clean Water Act objectives?

• Identify and discuss the key challenges to monitoring changes to wetlands, backwaters, and other riverine features along the Mississippi River corridor. How could these monitoring challenges be best addressed and overcome?

3. Water Quality Indicators and Standards:

• Identify and discuss the key challenges associated with establishing water quality indicators and standards in the 10 different states along the Mississippi River corridor. How could these processes be enhanced to ensure protection of downstream water quality? What benefits (if any) could be realized by making these procedures more uniform across the states?

• To what extent does the Clean Water Act affect water quality in the Gulf of Mexico? Do means exist to develop water quality standards—in particular nutrient standards—in all 10 Mississippi River states that can ensure protection of water quality in downstream waters in the Gulf of Mexico?

4. Policies and Implementation: Identify and discuss the key challenges in administering Clean Water Act authorities and programs aimed at

• National Pollution Discharge Elimination System (NPDES) permits,
• Impaired waters designations, and
• Protecting and restoring wetlands, backwaters, and other riverine features along the 10-state Mississippi River corridor.

How could collaborative efforts within federal agencies, between federal agencies, and between the 10 Mississippi River states, be strengthened to enhance implementation of these Clean Water Act provisions?

variably controlled municipal combined sewer and stormwater discharges, and steadily increasing industrial activities along the river, have contributed to water quality degradation. Concerns regarding deteriorating water quality in the nation's waterbodies were reflected in passage of both the Federal Water Pollution Control Act Amendments in 1972 and the further 1977 amendments that gave the act its shorter name, the Clean Water Act (also subsequently amended in 1981, 1987, and 1990). Since the Clean Water Act's passage, good progress has been made in many aspects of controlling "end-of-pipe" pollution from industries, municipalities, and other point sources to the Mississippi River. The focus of water quality improvement efforts generally has been on controlling these point sources through the act's National Pollutant Discharge Elimination System (NPDES), a permit program for pollutant discharges.

Point source pollution remains a problem in some stretches of the Mississippi River and in the river basin. There are also water quality problems related to "legacy" pollutants in the river, along with emerging water quality problems. The more pervasive Mississippi River water quality management challenge today is management of nonpoint sources of pollutants, especially nonpoint inputs of nutrients and sediments. These nutrients and sediments derive from a variety of activities, including agriculture, forestry, and increased urbanization. Mississippi River water quality also is affected by many large-scale hydrologic modifications along its mainstem, including the construction of locks, dams, and navigation pools, as well as levees along the lower river. A prominent example of the challenges posed by nonpoint source pollutants is the occurrence and persistence of a hypoxic (oxygen-deficient) "dead zone" in the northern Gulf of Mexico as a result of nutrient (nitrogen, in particular, but also phosphorus) inputs from the Mississippi River. Reductions of nitrogen and phosphorus in nonpoint source inputs to the Mississippi River are necessary to address this problem. Clearly there are important Mississippi River water quality issues beyond sediments and nutrients, such as point and nonpoint inputs of pathogenic microorganisms, toxic metals, and organic compounds, as well as the effects of legacy pollutants. Sediments and nutrients, however, are the factors of primary concern because of the magnitude of their mass loadings into the river, their changes over time, and the scale of the associated ecological impacts.

Under the Clean Water Act, the EPA and the states have joint responsibility for protecting, maintaining, and restoring water quality. In general, states designate specific uses for their waters; establish water quality criteria and, where required, maximum permissible discharge loadings to protect those uses; control pollutant sources through regulatory and nonregulatory measures; and monitor and assess water quality on an ongoing basis. States must submit periodic water quality assessment reports to the EPA, including lists of impaired waters, and then take appropriate actions to protect and restore impaired waters through the development of TMDLs (Total Maximum Daily Loads, which are discussed further in this report). The EPA has largely an oversight and coordination role in this process, establishing recommended water quality criteria and related elements of Clean Water Act administration (UMRBA, 2006). However, the EPA has authority to take action and lead when the objectives of the Clean Water Act are not being met by the states.

The 10 states along the Mississippi River corridor differ in the extent to which they have focused on water quality in the Mississippi River compared to other waters within their respective boundaries. Smaller streams and rivers contained entirely within state borders do not create interstate jurisdictional issues under the Clean Water Act. Responsibility for water quality monitoring and management in the Mississippi and other large, interstate rivers, however, is not well defined by the Clean Water Act or other legislation. Federal and state agencies conduct some water quality monitoring on the Mississippi River, but there is no comprehensive and systematic program or initiative designed to oversee Mississippi River water quality monitoring, protection, or restoration. Thus, despite the value and importance of the Mississippi River, there is no clearly defined, river-wide framework for adequately monitoring and ensuring protection of its water quality—a theme that runs through this report.

As specified in the charge to the committee, this report's objectives are to identify key water quality problems along the 10-state Mississippi River corridor, review the experience of Clean Water Act implementation along the corridor in addressing these problems, and assess whether and how the Clean Water Act framework can be used to address the problems more effectively in the future. Concerns regarding the management of nutrients and sediment loadings to the Mississippi River are key topics within this report, but a broad range of contaminants is also considered.

Other issues relevant to Mississippi River water quality management that may be important to readers of this report, but were beyond the scope of this study, include in-depth analysis of remediation of legacy contaminants across the river basin; valuation of environmental goods and services of the Mississippi River-floodplain ecosystems; possible amendments to the Clean Water Act; possible reallocations of federal and state resources devoted to water quality monitoring; operational changes of upper Mississippi River dams and navigation pools; environmental justice considerations that may relate to localized water quality conditions along the river; creation of new organizations for watershed and water quality management; efficiency of the existing TMDL process; and further analysis of cultural and historical differences among the 10 states along the Mississippi River corridor. While the importance of these and other issues is acknowledged, the committee focused its efforts and this report on the core Clean Water Act issues along the 10-state Mississippi River corridor as directed by its statement of task.

This report is focused on the Clean Water Act and its implementation. The committee was not specifically charged to consider possible statutory changes to the Clean Water Act. The committee discussed this topic and chose to conduct its investigations and present its findings and recommendations entirely within the framework of the existing act. Although the committee focused its report on CWA implementation, the findings and recommendations herein provide a foundation that could be used for water quality management and restoration activities in realms beyond the act.

REPORT ORGANIZATION AND AUDIENCE

Following this introductory chapter, Chapter 2 reviews the characteristics of the Mississippi River basin, with an emphasis on features and activities that affect water quality in the river and into the northern Gulf of Mexico. Chapter 3 presents an overview of the key provisions of the Clean Water Act and can serve as a primer on this topic. Chapter 4 examines key issues, advances, and challenges regarding implementation of the Clean Water Act along the 10-state Mississippi River corridor. Chapter 5 addresses logistical and administrative challenges in establishing and maintaining a water quality monitoring program on a large, interstate river such as the Mississippi. Chapter 6 discusses agricultural practices across the river basin and their implications for Mississippi River water quality. Chapter 7 explores institutional and policy modifications that could lead to more effective implementation of the Clean Water Act for the Mississippi River and other large rivers of the nation. The report's conclusions and recommendations are printed in boldface in the Summary, as well as in each summary section at the end of the individual chapters.

This report's target audience includes federal and state elected officials, federal and state resource managers and scientists, experts in river and water quality science and policy issues, nongovernmental organizations with interests in Mississippi River and northern Gulf of Mexico water quality, and individual citizens along the river, across the basin, and along the Gulf Coast. Environmental protection and agricultural agencies for states in the Mississippi River basin comprise a special audience for the report because the states have primary responsibility for implementation of the Clean Water Act and coordination with other states. The U.S. EPA and the U.S. Department of Agriculture, as well as the U.S. Geological Survey, constitute another special audience for this report, because leadership and coordination by these federal agencies will be crucial for more effective monitoring activities and for improving water quality in the Mississippi River and the northern Gulf of Mexico.

2

Characteristics of the Mississippi River System

The Mississippi River is one of the world's and the nation's great river systems. It ranks among the world's 10 largest rivers in size, discharge of water, and sediment load, and its drainage area covers 41 percent of the area of the conterminous 48 states (Milliman and Meade, 1983; Meade, 1995). With a length of roughly 2,300 miles, it is the second-longest river in the United States, exceeded in length only by the Missouri River (which is roughly 2,540 miles long and is the Mississippi's largest tributary). The Mississippi River watershed extends from the Appalachian Mountains in the east to the Rocky Mountains in the west, and from southern Canada southward to the Gulf of Mexico (Figure 2-1). The Mississippi's drainage area includes all or parts of 31 U.S. states; approximately 70 million people live in the basin. The Mississippi River enters the Gulf of Mexico through two deltas: the Mississippi River proper through its larger delta southeast of New Orleans, Louisiana, and the Atchafalaya River delta, located to the west on the central Louisiana coast.

The Mississippi River basin supports a high diversity and abundance of wildlife with their concomitant economic and social benefits. The Mississippi River valley is as an important international migration corridor for waterfowl and the site of the Upper Mississippi River National Wildlife and Fish Refuge, which is the longest river refuge in the continental United States. The river and its tributaries support a rich fish and invertebrate fauna, including several threatened and endangered species, such as the pallid sturgeon and several mussels. The Mississippi River, particularly in its upper reaches, has important commercial and recreational fisheries; the Upper Mississippi River National Wildlife and Fish Refuge hosts an estimated

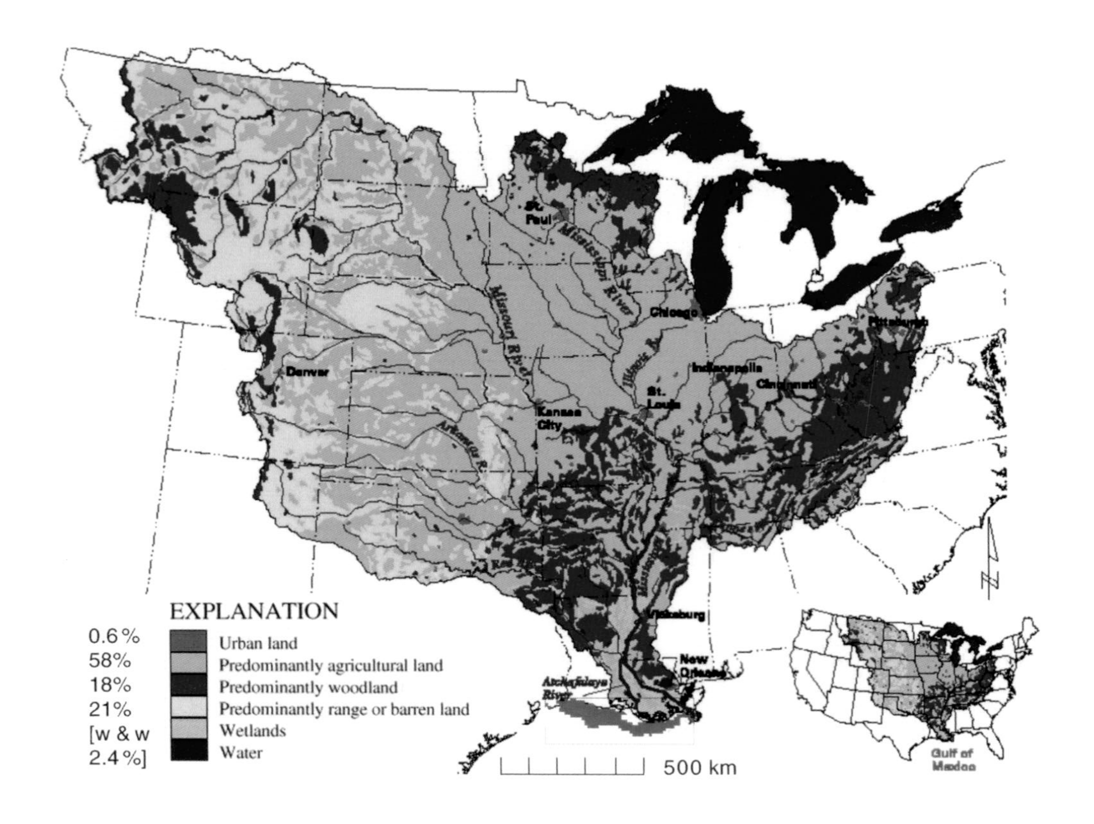

FIGURE 2-1 Mississippi River drainage basin, major tributaries, land uses, and the Gulf of Mexico hypoxic area (as of 1999).
SOURCE: Reprinted, with permission, from Goolsby (2000). © 2000 by the American Geophysical Union.

119 fish species (USFWS, 2007). Although the full economic values of these ecosystem assets and services may not be measured readily through market transactions, the economic impacts of recreation on the upper Mississippi River economy have been estimated at well over $1 billion (in 1990 dollars) annually (USACE, 1994).

In addition to these ecological resources, the Mississippi River serves as an important commercial transportation corridor. Hundreds of millions of tons of commodities are shipped annually on the Mississippi, and the river carries approximately 60 percent of the nation's corn exports and 45 percent of its soybean exports (USACE, 2004). Navigation on the upper river is supported by 29 locks and dams that impound a series of navigation pools, which have had substantial impacts on river ecology and biota.

The Mississippi River system's biotic resources and value for recreation and water supply depend on suitable water quality, which is affected by numerous factors and inputs across its vast river basin. The Mississippi River receives contaminants from both point (i.e., a specific site, such as effluent from a sewage treatment plant or an industrial site) and nonpoint (i.e., unconfined and often unregulated sources, such as cropland) sources. The Mississippi River thus exhibits various kinds of water quality degradation and changes in different reaches. The river's water quality is especially affected by nonpoint sources and, in particular, nutrient and sediment inputs (Meade, 1995; Howarth et al., 1996; Downing et al., 1999; Goolsby et al., 1999; NRC, 2000a; Figure 2-2). These nonpoint source pollutants derive from a variety of sources, including agricultural lands and city streets and yards. They also can be deposited on the landscape and surface waters from the atmosphere as a result of fossil fuel combustion and volatilization of ammonium from fertilizers and animal wastes. Applications of nitrogen and phosphorus fertilizers, primarily to row crops such as corn and soybeans, constitute the majority of nonpoint source pollutants (Howarth et al., 1996; Bennett et al., 2001; Turner and Rabalais, 2003; Figure 2-2).

This chapter presents an overview of the characteristics of the Mississippi River and its large and varied watershed, with an emphasis on features and land use in the watershed that influence Mississippi River water quality. As this chapter explains, the quality of water in the Mississippi River basin reflects both natural processes and human influences across varying scales of time and space. The chapter is divided into four sections: Mississippi River physiography and population; historic alterations of the Mississippi River system and its river basin; Mississippi River water quality; and water quality impacts on the northern Gulf of Mexico.

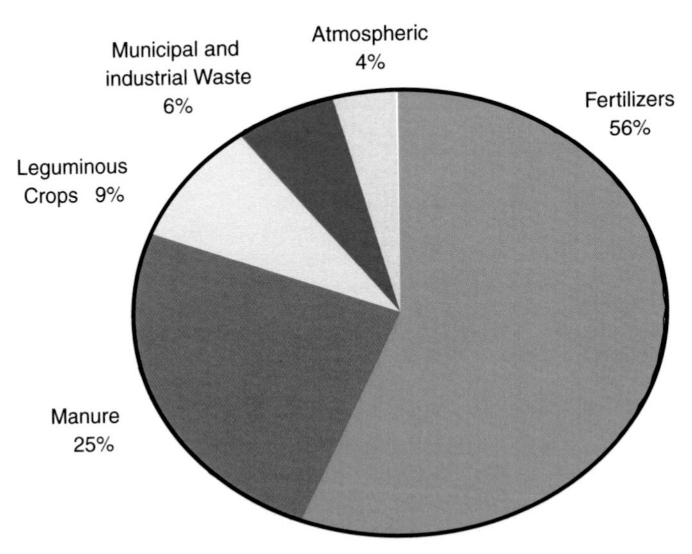

FIGURE 2-2 Relative proportions of point and nonpoint sources of nitrogen to the Mississippi River from the Mississippi River basin.
SOURCE: Based on Antweiler et al. (1995) and Goolsby et al. (1999).

THE MISSISSIPPI RIVER BASIN

Physiography and Population

Physiography

The Mississippi River system stretches from the river's headwaters at Lake Itasca in Minnesota southward through the heart of the continental United States, to the river's mouth at the Gulf of Mexico. The mainstem of the Mississippi River passes through or borders 10 states—Minnesota, Wisconsin, Iowa, Illinois, Missouri, Kentucky, Tennessee, Arkansas, Mississippi, and Louisiana. The Mississippi River is fed by several large tributary streams, including the Ohio River, the Missouri River, the Arkansas River, and the Red River. The Missouri River subbasin constitutes 42 percent of the Mississippi River basin area and dominates the Mississippi basin's land surface (Figure 2-3). Other major subbasins are the Ohio, Arkansas, and Red River subbasins, which comprise approximately 16, 13, and 7 percent of the entire river basin, respectively. The upper and lower Mississippi River basins comprise about 15 and 7 percent, respectively, of the surface land area that can affect Mississippi River water quality.

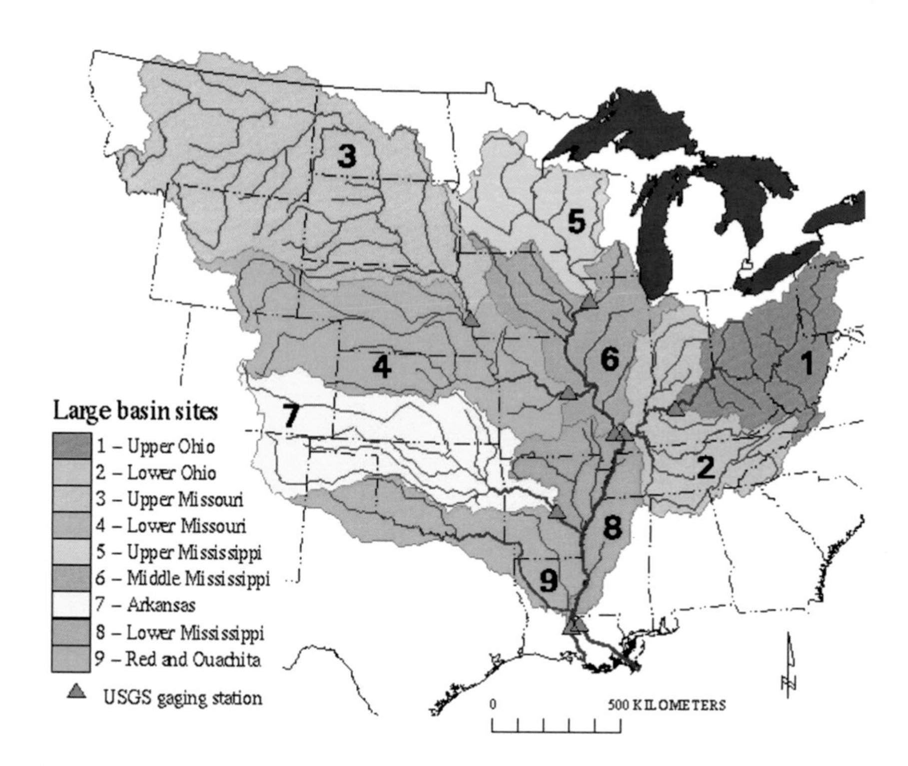

FIGURE 2-3 Major subbasins of the Mississippi River watershed.
SOURCE: Goolsby et al. (1999).

Landforms and landscape features affect runoff rates and the ability of the land to absorb water before it runs into waterways, both of which can affect water quality. Most of the Mississippi River basin is formed on low plateaus and the high plains (Hunt and Trimble, 1998). The eastern side of the basin borders the Appalachian Mountains, while the basin's western portions extend to the Continental Divide in the Rocky Mountains. Low plateaus across much of the basin generally are less than 1,000 feet in elevation, while the High Plains region in the Missouri and Arkansas watersheds ascends to the west and reaches elevations of 5,000 feet above sea level at the base of the Rockies.

The area north of the Ohio and Missouri Rivers was glaciated during the Pleistocene Era, and these landscapes are mostly flat to gently rolling ground moraines. Pleistocene glaciers left large areas of the midwestern United States, especially areas in Wisconsin and Minnesota, as wetlands and lakes. Over the past 150 years, many of the basin's wetlands and swamps—which have significant capacity to slow runoff and floodwaters and to trap and filter potential pollutants before they reach the Mississippi River—have been drained for agriculture (and, to a lesser extent, for urban development) and now are largely productive croplands (Prince, 1997). Natural wetlands and gentle slopes, however, do not characterize all of this area. For example, the "Driftless Area" between Red Wing, Minnesota, and Dubuque, Iowa, was not affected by at least the most recent stage of glaciations. Unlike the more subtle terrain of surrounding areas, the Driftless Area has picturesque bluffs, steep slopes, and local relief of several hundred feet. The region of the basin lying to the south of the Ohio and Missouri Rivers consists largely of unglaciated low plateaus, except for the broad Mississippi River valley below Cairo, Illinois. In many places—for example, along the river at Vicksburg, Mississippi—old coastal plain material is covered with alluvial deposits of the Mississippi River and its tributaries and with windblown loess from the upper Midwest deposited after the last ice sheets retreated some 10,000 years ago.

Approximately 60 percent of the river basin consists of agricultural land (Figure 2-1), and the central portion of the basin, extending from Iowa to Ohio and from the Ohio and Missouri Rivers northward almost to the Canadian border, supports extensive croplands. The area is generally flat to rolling with hot, wet summers having long days (i.e., >15 hours of daylight in many areas for much of the summer). This region is known as the "Corn Belt," but today it produces large amounts of both corn and soybeans. In the basin's more arid areas to the west, more drought-tolerant crops (e.g., wheat) are grown (see Fremling, 2005, for more detail on Mississippi River geology and landforms).

Population

Population distribution affects the different types and amounts of pollutants that reach the Mississippi River. For example, industrial point sources tend to be concentrated in cities, agricultural nonpoint sources tend to be in rural areas, and industrial sources tend to contribute more toxic pollutants than do rural areas. Population centers also are more likely to be the points of wastewater discharges. Given its large area, different parts of the Mississippi River basin have different—and sometimes widely disparate—population densities. Population density in the Mississippi River basin is approximately 6 people per square kilometer, which is relatively low in comparison to similar figures from, for example, the Chesapeake Bay watershed (90 people per square kilometer) or Long Island Sound (200 people per square kilometer).

Most (58 percent) of the basin's 71 million inhabitants live in cities or metropolitan areas with a population of 500,000 or more (U.S. Census Bureau, 2007). During 1990-2000, population in every Mississippi River basin state grew, and the region defined by the U.S. Census Bureau as the "Midwest" grew at a rate of 7.9 percent (U.S. Census Bureau, 2007).[1] Not all sections of the basin are experiencing population growth, however, and many rural counties in the basin are experiencing population declines. For example, most of the basin's population growth in the Midwest tends to be concentrated in its larger urban areas, such as Sioux Falls, South Dakota, and Minneapolis-St. Paul, Minnesota, which are growing rapidly.

Stresses on the Mississippi River system are affected by these differences in landforms and in human population within the river's subbasins, and by the capacity of receiving waters to dilute and otherwise reduce the effects of the specific types of pollutants generated in different locations. For example, some contaminants, such as fecal coliforms and some urban industrial toxic substances, are effectively diluted as they move downstream. Similarly, some toxic contaminants degrade or are sorbed to sediment and settle out. In contrast, other pollutants, such as some herbicides and pesticides, accumulate with distance downstream, either in the water itself or in the tissues of living organisms (Nowell et al., 1999).

Precipitation and Hydrology

The Mississippi River basin spans several climate zones, which affect the timing and amounts of rainfall (and pollutants) entering the river at various points along its path. The eastern and southern portions of the Mississippi River watershed generally receive more rainfall than the western and northern portions. Annual average values range from 60 inches or more in the southern Appalachian Mountains and along the Gulf Coast to 10-15 inches in the basin's westernmost portions. The northern half of the basin experiences a continental climate, with warm to hot summers and extremely cold winters, while the southern coastal region experiences a humid subtropical climate. In the north-central part of the basin, where agricultural activity is most intense, annual precipitation averages about 30 to 40 inches, with a pronounced summer maximum. Annual rates of evaporation vary greatly across the basin, ranging from 2-2.5 feet in the northeastern portions of the basin to as much as 5 feet in the southwestern part of the basin. The resulting annual runoff (precipitation minus actual evaporation) ranges from more than 20 inches in the east to less than 0.5 inch for much of the western part of the basin. The central agricultural region yields about 8-15 inches of runoff per year (Gebert et al., 1987).

Although the Missouri River watershed is roughly 2.5 times larger than the next largest of the Mississippi River's six major tributary watersheds

[1]Midwestern states listed in this category include Illinois, Indiana, Iowa, Kansas, Michigan, Minnesota, Missouri, Nebraska, North Dakota, Ohio, South Dakota, and Wisconsin.

(Figure 2-3), average annual Ohio River discharge is three times larger than that of the Missouri River. The Ohio River discharges more water into the Mississippi River than any of the river's major tributary streams (Figure 2-4 and Table 2-1). As illustrated in Table 2-1, the Ohio River watershed delivers 38 percent of the Mississippi River's flow, measured in terms of mean annual discharge. In comparison, the upper Mississippi River contributes 19 percent of the total of Mississippi River discharge into the Gulf of Mexico, followed by the Missouri River and the lower Mississippi River (13 percent each), the Arkansas River, and the Red River. Figure 2-4 illustrates the very different hydrologic character of the Mississippi River above and below Cairo, Illinois, which is located at the confluence of the Mississippi and the Ohio Rivers. The stark difference in upper and lower Mississippi River hydrology is important in the context of this study and is considered a crucial distinction throughout this report.

In addition to differences in discharge values and physical character across the river basin, Mississippi River flow varies seasonally and from

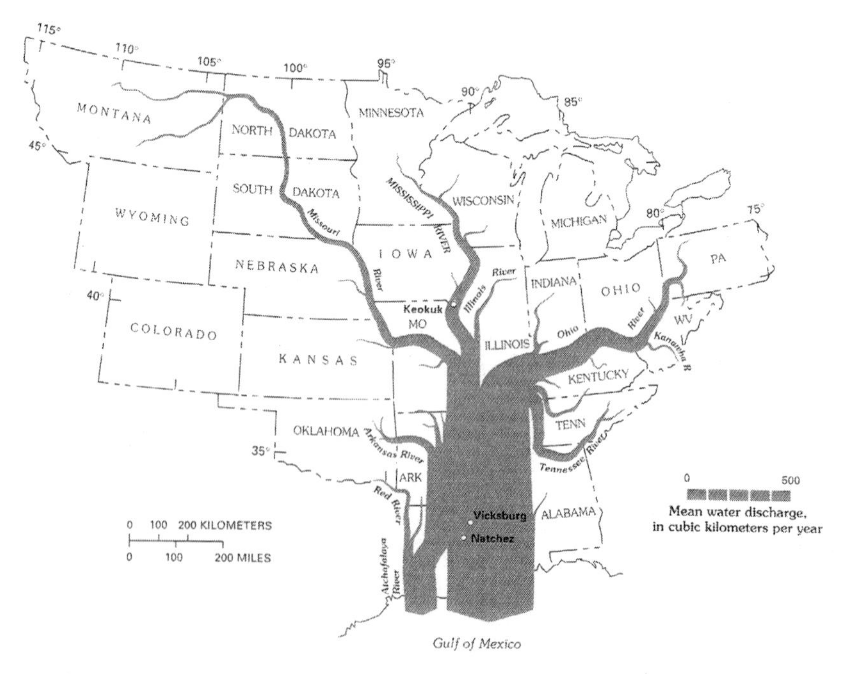

FIGURE 2-4 Relative freshwater discharge of Mississippi River tributaries to the amount delivered to the northern Gulf of Mexico. Widths of the river and its tributaries are exaggerated to indicate relative flow rates.
SOURCE: Meade (1995).

TABLE 2-1 Relative Proportions of the Mississippi River Watershed Within Its Larger Subbasins

Watershed	Land Area (%)	Discharge (%)
Upper (includes 5 and 6 in Figure 2-3)	15	19
Missouri	42	13
Ohio	16	38
Arkansas	13	10
Lower Mississippi	7	13
Red	7	7

SOURCE: Reprinted, with permission, from Turner and Rabalais (2004). © 2004 by Springer Netherlands.

year to year (Figure 2-5). In general, peak average flows—22,500 cubic meters per second—occur in March, April, and May, while low average flows—as little as 7,000 cubic meters per second—occur in late summer and early fall. The timing of water discharge affects the flux of materials from the basin's various landscapes. The timing, distribution, and temporal change of discharge volume into the northern Gulf of Mexico also affect both the physical oceanography and the biological processes leading to seasonal hypoxia (oxygen depletion). (See the section on Mississippi River water quality and the Gulf of Mexico for further discussion of hypoxia.)

HISTORIC ALTERATIONS OF THE MISSISSIPPI RIVER SYSTEM

Over the past two centuries, land use changes across the Mississippi River watershed and hydrologic changes along the length of its river-floodplain ecosystem have had significant impacts on water quality in both the Mississippi River and the Gulf of Mexico. One important land use change across the watershed has been substantial applications of nitrogen- and phosphorus-based fertilizers in the last half century, primarily to increase production of row crops. The region's land cover has changed dramatically, with vast areas of forests and prairies having been transformed into agricultural and urban lands. The river basin has also seen the drainage and conversion of millions of acres of wetlands, with more than one-half of the original wetland ecosystems having been converted to other land uses (Prince, 1997). Along the length of the river, key changes include the completion of a large hydropower dam at Keokuk, Iowa, in 1913; subsequent construction of locks, dams, and navigation pools as part of the 1930 Upper Mississippi River Navigation Project; and construction of flood protection levees along the entire river, especially in its lower reaches. This

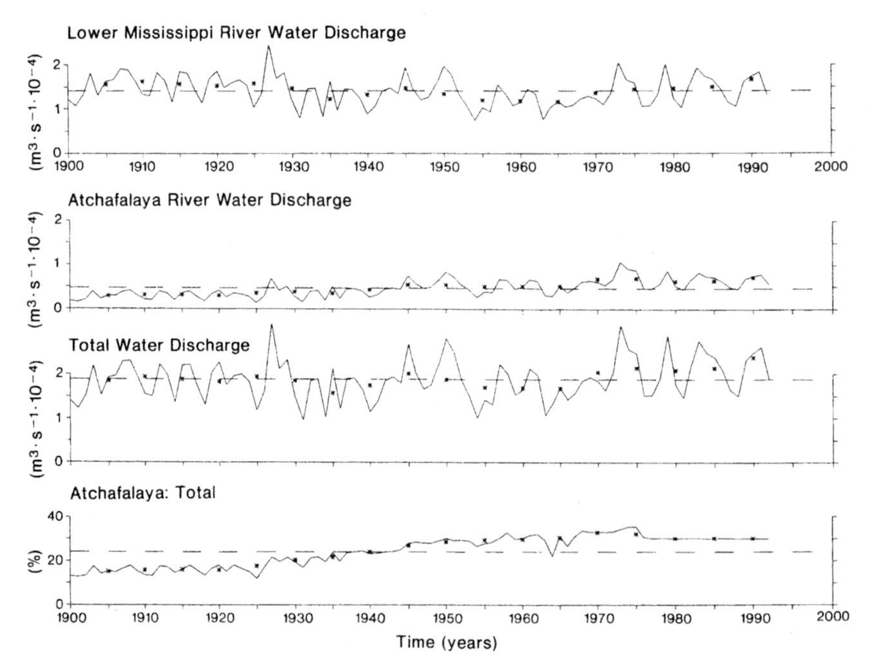

FIGURE 2-5 The 92-year annual average water discharge time-series data for the lower Mississippi River, Atchafalaya River, and combined flow. The lower panel shows the flow ratio (Atchafalaya River to total flow) for the same period.. Points are centered, decadal running-mean-averaged values (last values are partially extrapolated). Dashed horizontal lines are 92-year average values. Lower Mississippi River gauging station is located at Tarbert Landing, La. Atchafalaya River gauging station is located at Simmesport, La.
SOURCE: Reprinted, with permission, from Bratkovich et al. (1994). © 1994 by Estuarine Research Federation.

section discusses these changes as reflected in (1) land uses and wetlands, (2) navigation improvements on the upper Mississippi River, and (3) levee construction along the lower Mississippi River.

Land Uses and Wetlands

The conversion of vast areas of Mississippi River basin prairies and forests to cropland and other agricultural land following European settlement has had tremendous implications for Mississippi River water quality. Large areas of virgin forests across the basin had been cleared in the 1850s, and by 1920 they were reduced largely to remnant forests (Greeley,

1925). In the State of Ohio, for example, forest cover was reduced from 54 percent in 1853 to 18 percent in 1883 (Leue, 1886). The conversion of land to agriculture also inspires use of fertilizers and pesticides, which can become river pollutants.

The main use of land today in the Mississippi River basin is agriculture (58 percent of land use). Other land uses are range and barren land (21 percent), land types are woodland (18 percent), wetlands and water (2.4 percent), and urban land (0.6 percent; see Figure 2-1). Nevertheless, some forests are being reestablished today in parts of the river basin. Reversion of large areas of cropland in the eastern part of the basin since the 1920s has allowed regrowth of forest in part of the north-central region of the basin, some of which was in the "Prairie Archipelago" (Kuechler, 1975). Suppression of fire, reduced grazing, and expansion of land conservation by states and private organizations also have contributed to forest regrowth in certain areas.

Wetland ecosystems, once ubiquitous in the Mississippi River basin, serve important functions in regulating runoff and in reducing runoff of pollutants. Large losses of wetland areas, many of which were drained for conversion to agricultural land along the Mississippi River, have eliminated most of the natural buffering systems that could help reduce runoff of pollutants, toxic substances, and nutrients into the Mississippi River tributaries and mainstem (Table 2-2). Specifically, within the Mississippi River valley it is estimated that 56 percent of the wetlands have been lost to agriculture, navigation, reservoirs, and levees (Winger, 1986). Across the United States, similar rates of wetland losses have occurred. More than half of the original

TABLE 2-2 Wetland Losses in the Mississippi River Mainstem States

State	Percent Loss (circa 1980s)	Estimated Wetlands Remaining (acres)
Minnesota	42	8,700,000
Wisconsin	46	5,331,392
Iowa	89	421,900
Illinois	85	1,254,500
Missouri	87	643,000
Kentucky	81	300,000
Tennessee	59	787,000
Arkansas	72	2,763,600
Mississippi	59	4,067,000
Louisiana	46	8,784,200
Total	67	33,052,592

SOURCE: Dahl (1990).

wetlands in the United States have been lost to drainage practices (Zucker and Brown, 1998), many of which are related to agricultural production in areas that originally were swampy and too wet to farm.

Navigation Improvements on the Upper Mississippi River

On the upper Mississippi River, most changes to river hydrology and ecosystems have been driven by Congress and the efforts of the U.S. Army Corps of Engineers to improve river navigation. For example, the Rivers and Harbors Act of 1866 mandated a 4-foot navigation channel. In 1878, Congress authorized construction of a 4½-foot channel, which required the building of wing dams, closing of backwater channels, and building of five headwater dams to help control downstream flow (see also Anfinson, 2003). In 1906, Congress authorized a 6-foot channel that necessitated more and larger wing dams and additional closings of secondary channels crossing back swamp areas.

Despite this repeated channel deepening, the depth of the river channel for interstate commerce and transportation along the upper Mississippi River was not always dependable. Calls for a more reliable lock-and-dam system began in the late nineteenth century and increased in the early twentieth century. These discussions included some bitter controversies between navigation interests, on the one hand, and railroads and emerging environmental interests, on the other (Anfinson, 2003). After many years of discussion, Congress authorized the Upper Mississippi River Navigation Project in 1930. The Corps of Engineers subsequently constructed a system of locks, dams, and navigation pools to support a 9-foot channel, and by 1940 there were 27 low-head dams between St. Paul, Minnesota, and St. Louis, Missouri (Figure 2-6). The upper Mississippi River (along with the Illinois River, where several dams have been constructed and which is considered by the Corps of Engineers as part of the same navigation system—the Upper Mississippi River-Illinois Waterway—UMR-IWW—has promoted shipping and commerce in the region, as Mississippi River freight traffic increased from 2.4 million tons in 1940 to 87 million tons in 2000 (Anfinson, 2003). Lock-and-dam system proponents maintain that the UMR-IWW is essential to the competitiveness of commercial shipping, while some project critics emphasize the large environmental changes and impacts caused by the dams and navigation pools. From the late 1980s until 2004, the Corps of Engineers conducted a feasibility study of the economic prospects of extending several locks along the lower portion of the UMR-IWW. The study was the most extensive in the agency's history. It was completed in December 2004 when the final report recommended a $5.3 billion program for ecosystem restoration and a $2.4 billion program for navigation infrastructure improvements.

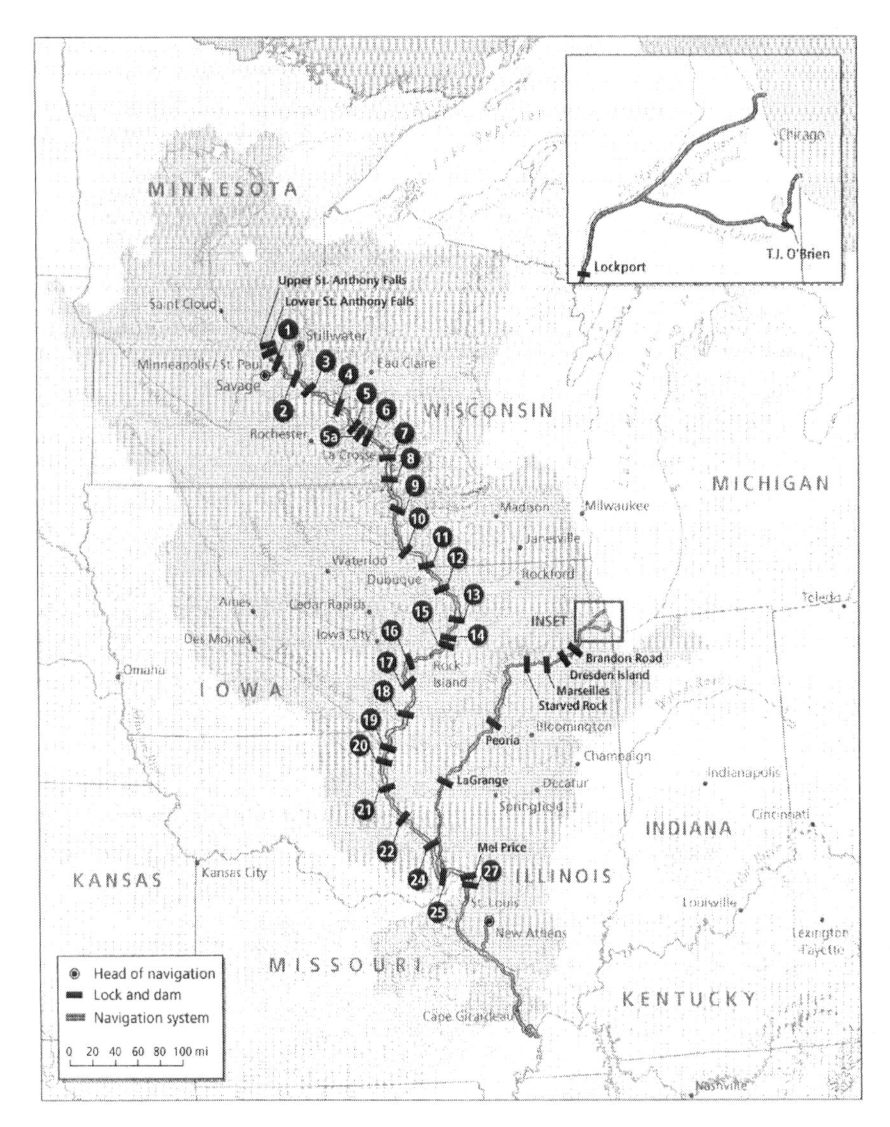

FIGURE 2-6 Locks and dams of the Upper Mississippi River-Illinois Waterway. SOURCE: NRC (2005).

Levee Construction Along the Lower Mississippi River

Large and extensive levees are the primary structures that affect flow and volume along the lower Mississippi River. There are no locks or dams across the Mississippi River below St. Louis. Levee construction in the lower Mississippi River began around 1717 and increased gradually until the 1880s, when the rate was accelerated (Barry, 1997). Following the disastrous 1927 flood in the lower Mississippi River region, the U.S. Army Corps of Engineers began an extensive flood control program of channelization and levee construction along the lower Mississippi, along with the establishment of floodways in Missouri and Louisiana. Levee construction has reduced considerably the natural floodplain of the Mississippi River basin and the natural aquatic ecosystems along its course (Table 2-3). Specifically, levees have reduced the area of seasonally flooded wetlands along the river, and dikes and revetments used to entrain the channel prevent the river from creating new habitat.

The reduced ability to form new habitats (which occurred historically as the river meandered) has had impacts on the floodplain, such as sedimentation of lakes on the lower Mississippi River in both oxbow lakes and other former channels (see Cooper and McHenry, 1989). In contrast to the upper Mississippi River, which has retained many of its larger backwater areas, fewer such backwater habitats remain along the lower Mississippi River.

Levees have not been the only source of hydrologic changes in the lower Mississippi River. During the 1930s, for example, the Corps of Engineers and others dug channels across the necks of meander loops, thereby shortening the river (Schumm and Winkley, 1994). The upper portion eventually filled with sediment, with the lower limb remaining as a link connecting the Mississippi, Atchafalaya, and Red Rivers. Eventually, the Atchafalaya River began enlarging itself through the capture of increasingly greater amounts of the Mississippi's flow. To prevent the Atchafalaya River from becoming the main channel of the Mississippi River, a series of control structures was completed in 1962 (Reuss, 1998; McPhee, 1999). Today, a controlled amount of Mississippi River discharge—roughly 25 percent—is diverted to the Atchafalaya system, joining the Red River, to the Gulf of Mexico (Turner et al., 2007).

TABLE 2-3 Losses of Floodplain Acreage Along the Mississippi River

River Segment	Floodplain Acreage × 1000	% of Floodplain Behind Levees
Upper Mississippi (N)	496	3
Upper Mississippi (S)	1,006	53
Middle Mississippi	663	82
Lower Mississippi	25,000	93
Deltaic Plain	3,000	96
Totals	30,493	90

SOURCE: Delaney and Craig (1997).

Effects of Structural Modifications

The hydrology of the vast Mississippi River basin system has been altered significantly by locks, dams, reservoirs and navigation pools, earthwork levees, channel straightening and bank stabilization, and spillways for purposes of flood protection, navigation, and water supply. These alterations have had numerous environmental impacts, including the transport and distribution of water, sediments, and dissolved materials (including nutrients and toxic substances), effects on the migration of fish and other aquatic species, submergence of aquatic vegetation, and the interruption of flow regimes. Large areas of the floodplain today are isolated by levees, the river is straightened and the flow is confined, large areas of floodplain in the upper river today are submerged under navigation pools, and many wetlands adjacent to the river have been drained. As a result, the spatial and temporal distributions of water velocities, bottom substrate, and water depths differ markedly from conditions that existed prior to the twentieth century.

MISSISSIPPI RIVER WATER QUALITY

Many Mississippi River water quality issues of today resemble the issues of the early 1970s, when the Clean Water Act was being drafted, but their relative importance has shifted in the past 35 years. Water pollution control measures (e.g., the National Pollutant Discharge Elimination System, discussed further in Chapter 3) have reduced point source pollutant inputs from industrial and municipal discharges. This has, in turn, reduced many serious water quality problems such as oxygen depletion caused by organic wastes, thermal pollution, oil slicks, phosphate detergent wastes, and sediments from larger construction sites. In addition, removal of lead from gasoline and the banning of some industrial chemicals such as polychlorinated biphenyls (PCBs) and pesticides such as chlordane, aldrin, dieldrin, and DDT (dichlorodiphenyltrichlorethane) have greatly reduced the amount of toxic substances in the Mississippi River. Pretreatment programs in larger cities have reduced discharges of heavy metals and other toxic materials from municipal wastewater treatment plants (see Chapter 4, Box 4-1 for further discussion of water quality improvements under Clean Water Act-related projects).

Despite these advances, the Mississippi River today is affected by water quality problems and challenges that include nutrients, sediments, toxics, and fecal bacteria. Toxic substances—metals and organic chemicals—are primarily legacy contamination issues, although there are continuing inputs, especially of pesticides. These substances have chronic ecosystem and human health impacts and are difficult to address, because river bottom sediments are the primary reservoir and source of these materials in many reaches of the river. High counts of fecal bacteria, once a public health problem at raw sewage discharges all along the Mississippi River, were substantially reduced with the implementation of secondary sewage treatment in many areas. Today, some parts of the river—mainly near large municipalities—still experience fecal bacteria counts that exceed water quality standards.

Fecal bacteria and new inputs of toxic substances can be controlled through existing mechanisms in the Clean Water Act. By contrast, water quality problems related to nonpoint source inputs—especially (1) nutrients, primarily nitrogen and phosphorus from agricultural runoff and other agriculture activities, and (2) sediments, from upland or farmland erosion and river bed and bank erosion—are not as readily addressed by existing mechanisms. Accordingly, this report focuses primarily on Mississippi River water quality problems as they relate to nutrients and sediments.

Nutrients

Excess nutrient loadings cause marine algae to grow to great abundance and thereby affect coastal aquatic ecosystems, both in the Gulf of Mexico and around the world. The processes of algae decomposition ultimately lead to oxygen depletion and "dead zones" in coastal waters. The Gulf of Mexico is probably the best-known of these affected coastal ecosystems, but nutrient overenrichment affects coastal areas both elsewhere in the United States (e.g., Chesapeake Bay) and in sections of Asia, Europe, and South America. Moreover, according to a recent report from the United Nations Environmental Programme (UNEP, 2006), the number of these dead zones is increasing. These consequences stem from global human population growth and its associated activities that have, at accelerating rates, altered the landscape, hydrologic cycles, and flux of nutrients essential to plant growth, particularly in the last half-century (Vitousek et al., 1997; Galloway and Cowling, 2002; Galloway et al., 2003). To support the need for fuel, fiber, and food, humans have increased nitrogen and phosphorus loadings to aquatic and terrestrial ecosystems significantly, altering the global cycles of those nutrients.

The excess nutrients affecting the Mississippi River-northern Gulf aquatic system derive primarily from diffuse, nonpoint sources (e.g., land runoff and atmospheric deposition) and stimulate a variety of ecological and related effects. This section discusses the role of nutrients in phytoplankton growth, nutrient quantities and changes over time, sources of nutrients within the Mississippi River basin, and the effects of excess nutrient loading.

Phytoplankton Growth

Plants of all types, including corn, soybeans, wheat, aquatic vegetation, seaweed, and microscopic phytoplankton or algae, need nitrogen and phosphorus to grow. Crop plant growth will not continue or reach maximum productivity without adequate nutrients, and farmers generally use nitrogen- and phosphorus-containing fertilizers to supplement nutrients in the soil. Similarly, aquatic plants (including phytoplankton) will not grow without suitable dissolved nutrient supplies. A specific group of phytoplankton that is the base of many aquatic food webs, the diatoms, also requires silica, which is essential for the formation of their cell walls. As with crops, the addition of nutrients to ambient waters stimulates phytoplankton growth. However, once nutrient loads cause aquatic systems to cross certain thresholds, the results are not entirely positive. Instead, excess nutrients can reduce water clarity and stimulate harmful algal blooms. They can also lead to oxygen depletion, which in some cases can cause reduced or lost fisheries production.

Naturally low availability of nitrogen, phosphorus, or silica, either in absolute concentration or in relation to other nutrients, may limit phytoplankton growth. As a result, introducing excess supply of the limiting nutrient will enhance phytoplankton growth. Phosphorus usually is considered the limiting nutrient for phytoplankton growth in freshwater systems and nitrogen in marine systems (Rabalais, 2002), but other, perhaps multiple, nutrients may be limiting.

Thus, both the concentration of a nutrient and its abundance relative to other nutrients control the production and composition of phytoplankton. Excess phosphorus has caused notable water quality problems in freshwater systems, such as noxious and toxic algal blooms, decreased water clarity, and low dissolved oxygen conditions. Likewise, excess nitrogen and sometimes phosphorus have led to algal blooms in estuarine and coastal marine systems with the same results.

Nutrient Quantities

The amount of nutrients in an aquatic system can be quantified by concentration, loading on and/or yield from a landscape in a watershed, and loading to waterbodies. Concentrations of silicate and various forms of nitrogen and phosphorus have been measured frequently since the 1950s at many locations in the Mississippi River and near the terminus of the Mississippi, both at St. Francisville and at New Orleans. Earlier data are available from the twentieth century, from 1905-1906 and from 1935-1936. Data from the lower Mississippi River show that the average annual nitrate concentration rose from the 1960s through the early 1980s, and considerably more since the end of the twentieth century (Turner et al., 1998; Goolsby et al., 1999). Similar changes are seen throughout the Mississippi River basin (Figure 2-7).

Nitrate load in the Mississippi River (the product of nitrate concentration × discharge) increased about 300 percent from the 1950s to the mid-1990s (Goolsby et al., 1999; Goolsby and Battaglin, 2001), whereas streamflow from the basin increased only 30 percent in the same period (Figure 2-8 and Bratkovich et al., 1994). Clearly, the most significant driver

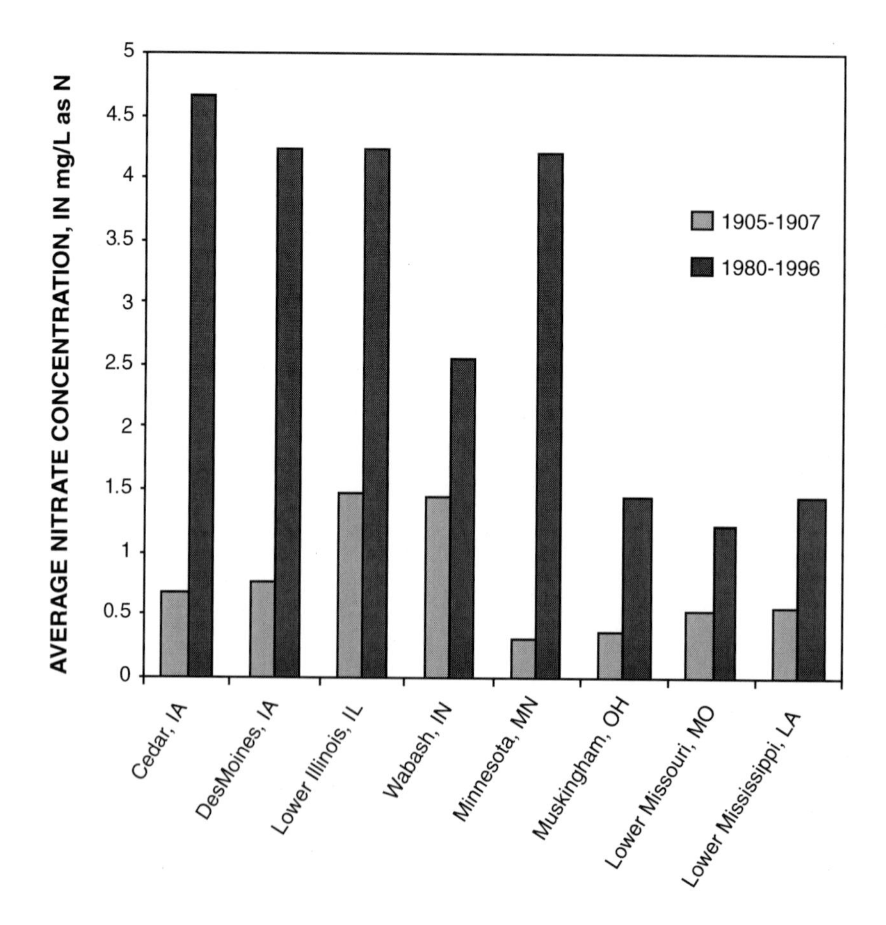

FIGURE 2-7 Comparison of average annual nitrate concentrations in 1905-1907 with those in 1980-1996 for the Mississippi River mainstem and some of its tributaries.
SOURCE: Reprinted, with permission from Goolsby (2000). © 2000 by American Geophysical Union.

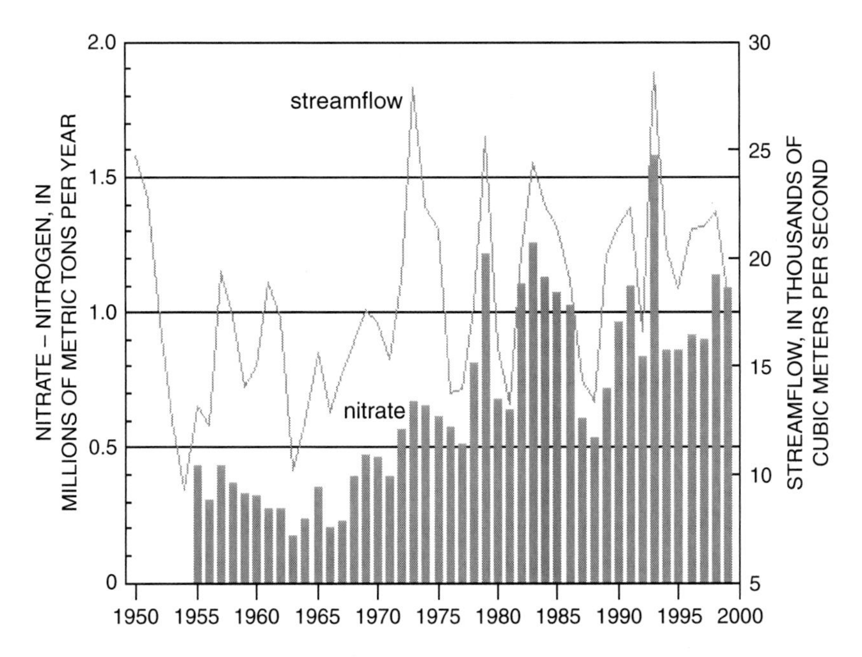

FIGURE 2-8 Annual flux of nitrate from the Mississippi River basin to the Gulf of Mexico, 1955-1999, and mean annual streamflow, 1950-1999.
SOURCE: Reprinted, with permission from Goolsby (2000). © 2000 by American Geophysical Union.

of the change in Mississippi River nitrate load is the increase in nitrate concentration, not freshwater discharge (Justić et al., 2002). Only 20 to 25 percent of the increased nitrate load between the mid-1960s and the mid-1990s was attributable to greater runoff and river discharge, with the remainder due to increased nitrogen concentrations in the lower river (Donner et al., 2002; Justić et al., 2002). River discharge is governed by precipitation (less evapotranspiration) that can be regulated only marginally by dams on the Mississippi River mainstem and its tributaries, because the mainstem upper Mississippi River dams do not create reservoirs, but navigation pools. Thus, nitrate loadings to the Mississippi River can be controlled effectively only through control of nitrate concentrations.

The total nutrient discharge to the Gulf of Mexico from the Mississippi River is dominated by nitrogen, with a mass loading that is about an order of magnitude greater than the phosphorus loading. From 1980 to 2005, nitrogen loadings ranged from 0.8 million to 2.2 million metric tons per year. Over the same period, values of phosphorus loadings were between 0.08 million and 0.18 million metric tons per year (Aulenbach et al., 2007).

Thus, nitrogen is the primary nutrient of concern in the northern Gulf of Mexico and along much of the Mississippi River. Excess phosphorus is a concern in various Mississippi River backwaters and tributaries and has significant impacts in certain sections of the Mississippi mainstem. A good example is at Lake Pepin, on the mainstem Mississippi River in southern Minnesota. Algal blooms and other impacts from phosphorus loadings occur there, especially at low flows in the river, and a Total Maximum Daily Load (TMDL) for phosphorus is in development (MPCA, 2007). Phosphorus sometimes is important in the lower Mississippi, where it can be a limiting nutrient to phytoplankton growth in the spring, and in the immediate plume of the Mississippi River as it discharges to the northern Gulf of Mexico. Given the importance of both nitrogen and phosphorus in various forms, it is necessary to consider management of both of these nutrient inputs, which stem primarily from nonpoint sources.

Nutrient Sources

Nutrients reach waterways through several pathways—erosion of nutrient-bearing soils and sediments, natural dissolution from soils and sediments, runoff over land or through soils, atmospheric deposition, and point source discharges. Nitrogen can be converted from atmospheric gas to ammonia and nitrate by bacteria on the roots of leguminous plants. Nutrients in the Mississippi River basin originate from the same multiple sources, but mostly from diffuse nonpoint sources (Figure 2-9). Figure 2.9 shows mineralization of soil organic nitrogen as an estimated constant background input rate (based on an assumption of 3 percent nitrogen in soil organic material and mineralization at a constant annual rate of 2 percent). Figure 2-9 also shows that inputs of nitrogen from agricultural sources, especially fertilizer applications, have been increasing and now are equal in magnitude to the natural background input rate. About 90 percent of the nitrogen load reaching the Gulf of Mexico from the Mississippi River is from nonpoint sources, including approximately 58 percent from fertilizer and mineralized soil nitrogen. The remaining approximately 10 percent is from a mix of sources that includes primarily municipal and industrial point sources (Goolsby et al., 1999; see Figure 2-9).

Most nutrients derived from the Mississippi River watershed are from its upper and middle portions (Goolsby et al., 1999). The dominant watershed in terms of total nitrogen loading is the combined upper and middle Mississippi watershed (subbasins 5 and 6 in Figure 2-10), with contributions of 35-45 percent, followed by inputs from the Ohio watershed at 28-30 percent. The total nitrogen and nitrate loadings from the Red and Arkansas River watersheds are relatively small compared to the others (less than 7 percent, each). Loadings of total phosphorus (Figure 2-10) and sili-

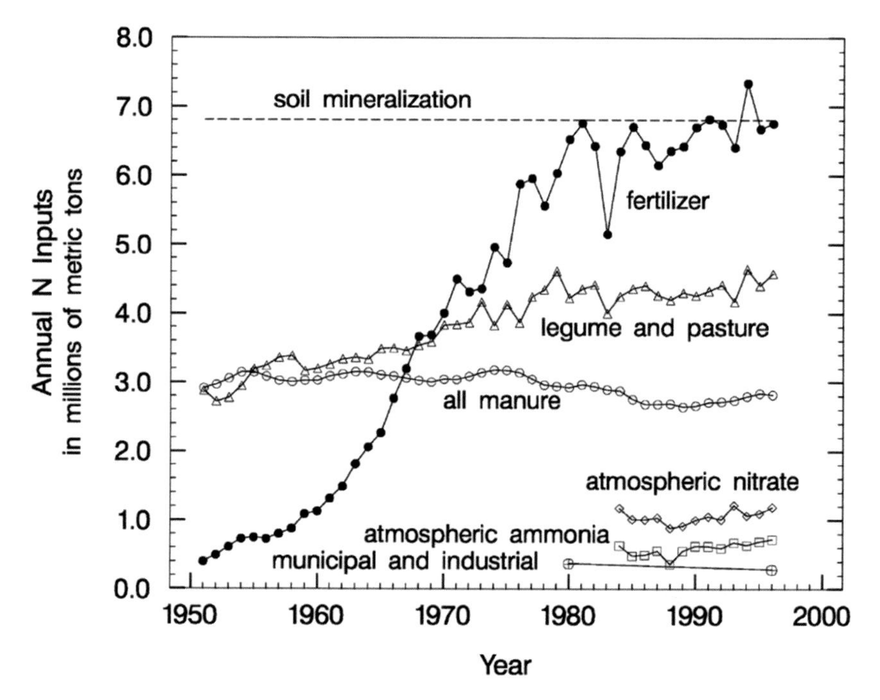

FIGURE 2-9 Annual nitrogen inputs from major sources in the Mississippi River basin, 1951-1996. Details of sources of data and methods for estimating inputs are in Goolsby et al. (1999).
SOURCE: Goolsby et al. (1999).

cate (not illustrated) are about equally divided among the combined upper and middle Mississippi, lower Mississippi, Ohio, and Missouri watersheds and are relatively high. The majority of the nitrogen and phosphorus flux—for example, 56 percent of the nitrate—is from above the confluence of the Ohio River with the Mississippi River, and it derives mainly from nonpoint agricultural sources (Goolsby et al., 1999; Turner and Rabalais, 2004).

Atmospheric deposition contributes a small (approximately 10 percent) percentage of nitrogen loading to the Mississippi River (Figure 2-9). The highest levels of nitrate deposition—which results from the burning of fossil fuels—are in the upper to middle Ohio River basin (Figure 2-11). The deposition of ammonium is highest within the upper to middle Mississippi River basin and is attributed to the volatilization of ammonia from fertilizers and animal wastes (Figure 2-11). Given the airborne nature of this pollutant, it may be more appropriately managed through the Clean Air

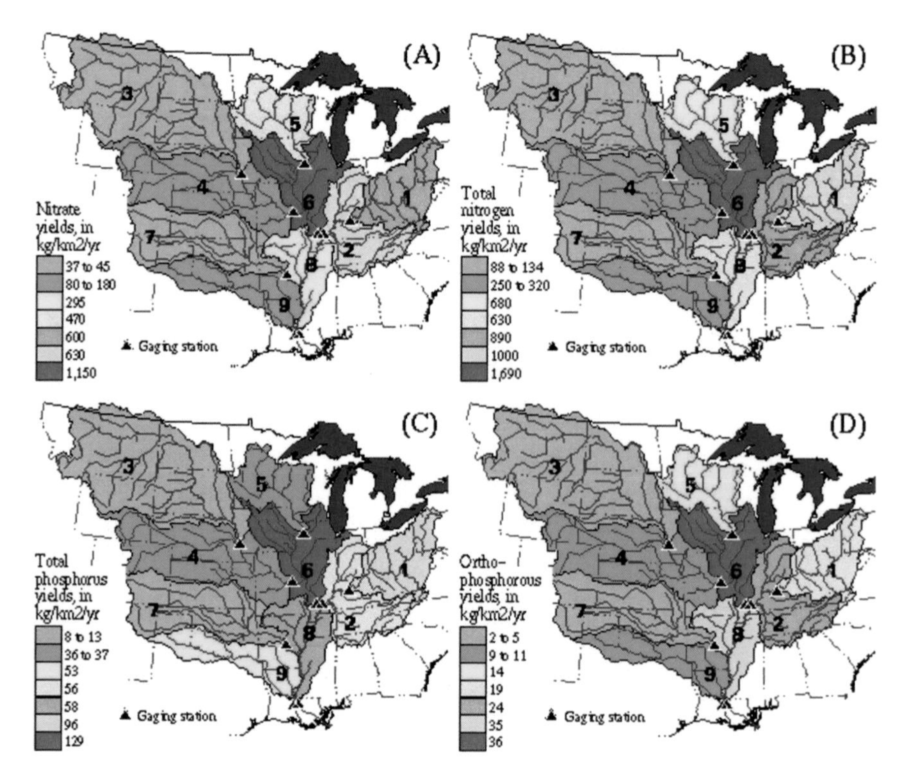

FIGURE 2-10 Spatial distribution of the average nutrient yields in nine large basins during 1980-1996.
SOURCE: Modified from Figure 4.5 in Goolsby et al. (1999).

Act. This deposition, however, eventually contributes to the total loading of nitrogen that may have to be managed under provisions of the Clean Water Act. A coordinated effort to manage the nutrient content of the Mississippi River needs to account for the multiple sources of nutrients that affect water quality and the activities that generate them.

Nutrient Uptake and Transformation

The proximity of sources to large streams and rivers is an important determinant of nitrogen delivery to coastal waters receiving Mississippi River discharge. The uptake and removal of nitrogen in the smaller streams is greater than in the Mississippi River mainstem, where this rate may ap-

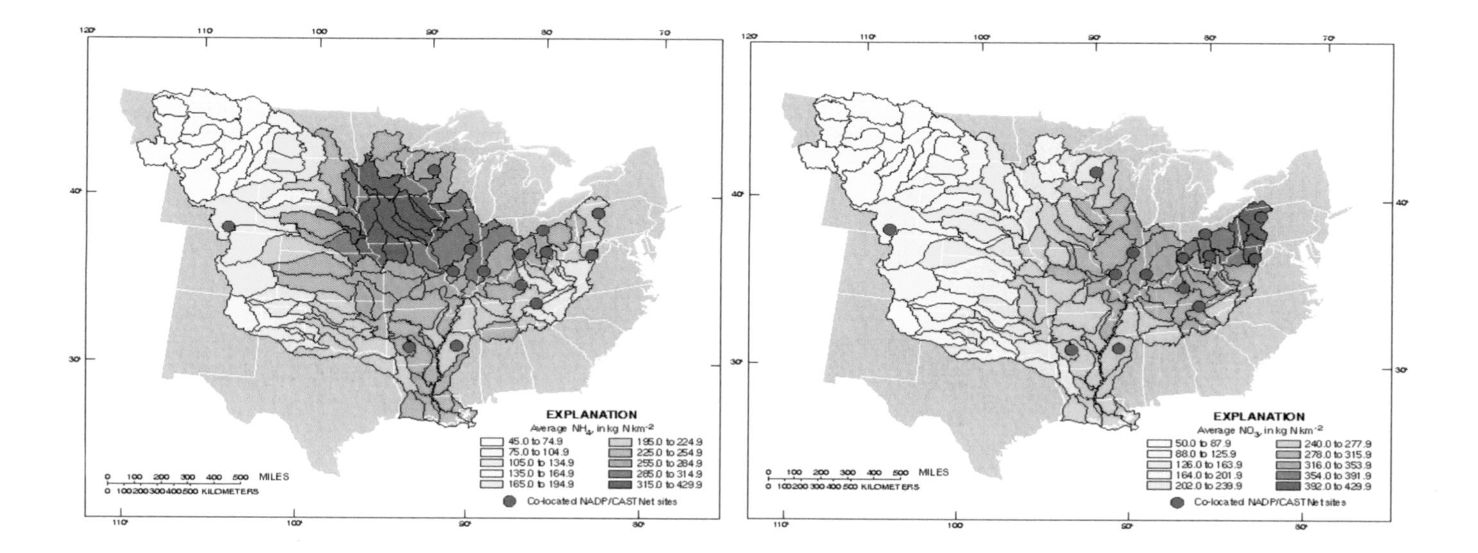

FIGURE 2-11 Wet deposition of NH_4^+ and NO_3^- averaged for 1990-1996 data from the National Atmospheric Deposition Program in each of the 133 accounting units that make up the Mississippi River basin.
NOTE: Blue circles indicate where NADP and CASTNet (Clean Air Status and Trends Network) sites are co-located.
SOURCE: Goolsby et al. (1999).

proach zero (Alexander et al., 2000). The entry location of nutrient loads to the Mississippi River system thus determines the relative influence of in-stream removal of nutrients through natural processes. Efforts to remove or reduce nutrients through management scenarios generally are more effective at the source of nutrient loads in smaller streams and rivers, rather than attempting to recover or mitigate nutrient loads once the nutrients enter the Mississippi River.

This is not to say that river floodplain projects designed to help remove nutrients are not valuable. In fact, nutrient removal projects downstream may in some instances be easier to implement than in upstream reaches for a variety of reasons, including financial, administrative, and others. Some of these "nutrient farms" are being planned in floodplain areas along the Illinois River, for example, which is a large Mississippi River tributary that delivers significant amounts of nutrients into the Mississippi. The Metropolitan Water Reclamation District of Greater Chicago is providing financial support for these projects, which are located approximately 100 miles downstream of its sewage outfalls in Chicago. River water is diverted through a series of gated, shallow floodplain compartments and then returned to the river, with a goal of demonstrating that substantial nitrogen removal can be achieved at less cost than with proposed tertiary treatment sewage plants in Chicago (Hey et al., 2005a, 2005b).

Existing and restored floodplains along the Mississippi and its major tributaries may reduce nitrogen loads to the Gulf during major flood events (25- to 500-year flood events) when floodplains are inundated. The "pulse" of nutrients delivered to the Gulf of Mexico during these events thus may be lessened. After the flood of 1993 in the upper Mississippi River basin, nitrogen and phosphorus in sediment deposits on the floodplains of the Mississippi and Illinois Rivers were well in excess of the growth requirements of floodplain vegetation and represent nutrients that were trapped instead of delivered to the Gulf of Mexico (Sparks and Spink, 1998; Spink et al., 1998).

In general, increased nitrogen loads to the Mississippi River are less likely to be taken up and transformed across the current Mississippi River basin than they were historically because of losses of the system's natural assimilative capacity. The human-modified landscape and hydrology of the Mississippi River system over centuries, coupled with population growth, agriculture, industrialization, urbanization, increased combustion of fossil fuel, and increased use of fertilizers in the post-World War II era, all have reduced the capacity to remove contaminants naturally across the entire watershed. Today's significant water quality problems in the Mississippi River basin and its offshore coastal waters are related to these landscape developments, coupled with increased nutrient loads derived primarily from agricultural fertilizers and activities.

Effects of Excess Nutrients

The loading of lakes, rivers, and coastal waters with previously scarce nutrients, such as nitrogen or phosphorus, usually boosts production of phytoplankton. In excess, these algae are linked to a number of problems in aquatic ecosystems, including murkier water, unpleasant odors and sights of decomposing algae, production of toxic substances, periods of oxygen depletion, and loss of important fisheries. High levels of phosphorus can degrade inland waters; turn pristine, clear lakes into weed-choked waterbodies; and accelerate bog succession. Excess levels of nitrogen seeping into groundwater can contaminate drinking water wells and supplies (see, for example, Burkholder et al., 1999; Gilbert et al., 2005).

Elevated nitrate levels in drinking water have serious public health implications. They are especially dangerous for children under 6 months of age because nitrate robs their blood of oxygen and can cause "blue-baby" syndrome. Removing nitrate from drinking water supplies is also an expensive proposition, requiring the addition of denitrification treatment systems. Elevated nitrate levels have been a problem in some areas of the Mississippi River basin. For example, in the Ohio River watershed, water quality advisories are issued every spring in Columbus, Ohio, for excess nitrate levels in local waters (Mitsch et al., 2001).

Excess nutrients in lakes, ponds, slow-moving streams, and brackish areas in the upper ends of estuaries often lead to blooms of cyanobacteria (blue-green algae) that produce toxic substances. Exposure of humans to these toxic substances through contact, inhalation of water spray, or oral ingestion can cause debilitating illness and even death. Recreational activities such as swimming and water skiing can result in exposure to contaminated water, as can being on the water in recreational or commercial fishing. Little is known about the transfer of cyanobacterial toxins into the food web, but recent studies indicate that there may be both environmental effects and human health concerns (Rabalais, 2005).

Sediments

The functioning of natural backwater and floodplain ecosystems along the Mississippi River depends on delivery of sediment and nutrients during floods. At the same time, sediment regularly fills some channels and other deep areas of the system and must be removed to support recreation and navigation activities and to sustain wildlife habitat. A multitude of contaminants (e.g., phosphorus, pesticides, heavy metals, PCBs) are often adsorbed onto or otherwise associated with sediment particles. Thus, many areas in the river system where sediment is deposited can become "hot spots" for a mix of plant nutrients and toxic substances.

Sediments from natural erosion, agricultural land loss, and bed and bank erosion that are suspended within the water column decrease water clarity, which often leads to water quality impairments. Soil erosion is also problematic because soil nutrients (especially phosphorus) and pesticides may be adsorbed onto soil particles and thus have the potential to pollute downslope or downstream. Furthermore, suspended sediments can become trapped behind dams and other engineered structures throughout the Mississippi River basin. The results are (1) sedimentation and trapping of sediments in areas such as navigational pools and backwaters on the upper Mississippi River and within the Atchafalaya River basin and (2) sediment deprivation in the Mississippi River deltaic plain, where combined natural and human-caused factors are leading to loss of coastal wetlands and barrier islands. Sediment-related problems along the Mississippi River thus range from too much to not enough sediment in different sections along the length of the Mississippi River corridor and can result in impairments to ecosystems and water quality.

Figure 2-12 illustrates relative sediment contributions from the Mississippi River and its main tributary streams (Meade, 1995). It is estimated

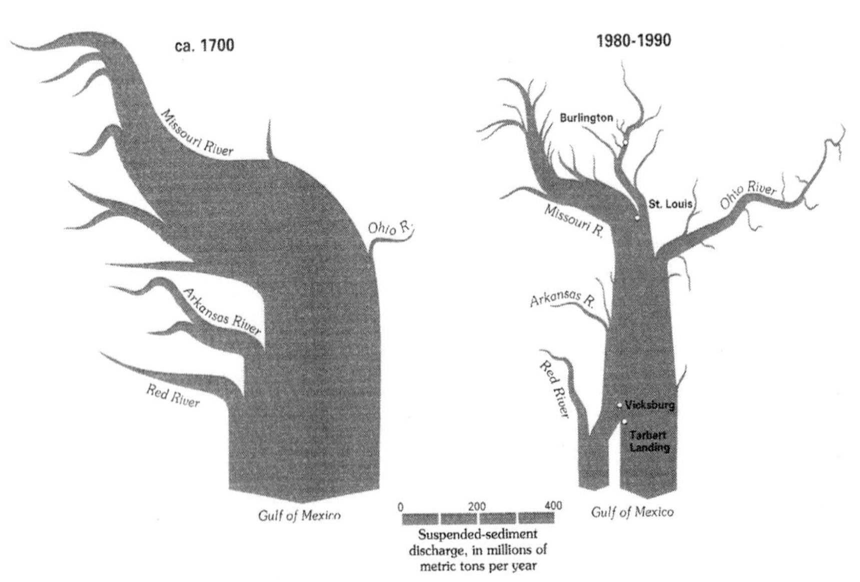

FIGURE 2-12 Mississippi River suspended sediment discharge, around 1700 (estimated) and 1980-1990. Values in millions of metric tons per year. Widths of the river and its tributaries are exaggerated to reflect relative sediment loads.
SOURCE: Meade (1995).

that the load of suspended sediments to the Gulf of Mexico in 1700 was roughly double the average values from 1980 to 1999 and that the Missouri River clearly dominated that load (Meade, 1995). Under present conditions the Missouri River continues to dominate the load, but because of the construction of several large storage dam reservoirs on the Missouri in the 1950s and 1960s that capture sediment, and because of land use changes in the upper Mississippi and Ohio River valleys, contributions of the upper Mississippi and the Ohio Rivers are proportionally greater than they were in the 1700s (Meade, 1995).

Figure 2-13 presents annual average estimates of Mississippi River suspended sediment loads at New Orleans for much of the twentieth century. The figure shows a steadily declining trend of suspended sediments in the river. It should be noted that estimates of suspended sediment yields and loads from the Mississippi River watershed vary among investigators because of variability in water discharge; length and completeness of the period of record; effects of variations of velocities with depth; logistical issues related to working in a large river; and sampling frequency (Meade, 1995; Turner and Rabalais, 2004).

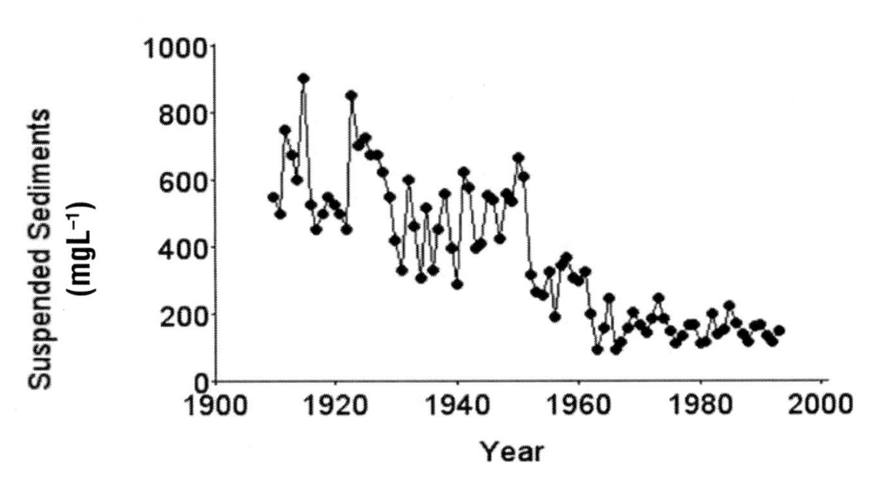

FIGURE 2-13 Annual average suspended sediment concentrations in the Mississippi River at New Orleans, Louisiana.
SOURCE: Reprinted, with permission, from data compiled by R. E. Turner, Louisiana State University. Data from New Orleans Sewage Board at Carrolton Treatment Plant, 1909-1993. Annual averages represent at least weekly, if not daily, measurements. All measurements were from gravimetric methods.

As shown in Figure 2-12, sediment inputs to the lower Mississippi River historically have been dominated by Missouri River flows. Although present and historical suspended sediment data have some inconsistencies, the data starting about 1915-1920 began a more reliable period of suspended sediment measurements for the system and heralded a period of decline in suspended sediment. Dams for flood protection and to enhance navigation were constructed on the Ohio River and in the upper Mississippi basin, respectively, in the 1930s. Large dams also were constructed in the Tennessee basin (Wilson Dam at Muscle Shoals was completed in the 1920s; others followed in the 1930s after the Tennessee Valley Authority was established in 1933) and on the Missouri River (1950s and 1960s). The decrease in suspended sediments has occurred mostly since 1950, when the largest natural sources of sediments in the drainage basin were cut off from the Mississippi River mainstem by the construction of large storage dams on the Missouri and Arkansas Rivers (Meade and Parker, 1985; NRC, 2002). These dams trapped large amounts of sediments and altered transport patterns of suspended sediments downstream in the basin all the way to New Orleans and into the Gulf of Mexico (Meade et al., 1990).

Present downstream sediment loads, however, may often be derived in considerable part from stream channel and bank erosion (Trimble, 1977, 1999). Sediment particles from current erosion may go into storage at the base of slopes or on downstream floodplains. Conversely, erosion of stream banks and channels may entrain sediment that may be anywhere from a few hours to a few millennia old. This situation is especially important in stream basins that have suffered heavy soil erosion in historical time, where many legacy sediments exist. Thus, downstream sediment yields may not reflect the quantity or the quality of material being currently eroded from slopes (Glanz, 1999; Trimble, 1976, 1977). As a result, current efforts to reduce soil erosion may be successful, but results often are not measurable downstream, at least in the short run.

The effects of sediments and sedimentation on water quality and habitat are important issues along the upper Mississippi River and provide examples of excess sediment as a pollutant. The Minnesota River, for example, contributes a large amount of sediment to the Mississippi just below Minneapolis-St. Paul. The Minnesota River runs through agricultural land in southern Minnesota and transports large loads of nutrients, pathogens, pesticides, and sediments. A large portion of the Minnesota River sediments delivered to the upper Mississippi River is deposited in Lake Pepin (less than 50 kilometers downstream), resulting in the gradual filling of that large, natural river impoundment.

In contrast to upper parts of the Mississippi River watershed where large amounts of sediments are input to the river and its floodplain and backwater areas, the deltaic plain of the Mississippi River is receiving less sediment than it did historically. The 9,600 square mile deltaic plain was formed and sustained over the last 6,000 years by delta lobe switching, crevasses, river floods, storms, tides, and wetland plants (Penland et al., 1988). Wetlands across the coast survived for centuries after they received substantial inputs of river sediments. Before the construction of the numerous large levees along the lower Mississippi River, regular overbank flooding and crevasses maintained the river's sediment input to the coastal landscape. The extensive tidal wetlands and other landforms of the Louisiana coast rapidly deteriorated with increased river control, particularly during the last half of the twentieth century. These changes have been driven by a variety of activities. The closing of distributary channels and construction of artificial levees along the river limit the nourishment of wetlands with sediments and fresh water. The numerous canals across the Mississippi River delta that have been dredged for navigation, oil and gas production, and transportation, have caused widespread hydrological modifications. The delta region also experiences relatively high rates of subsidence; the reduction in sediments that could help compensate for these losses has contributed to the deterioration of barrier islands along the Louisiana coast (Boesch et al., 1994). Although some of these processes are natural, most of these environmental changes have been due to human activities that have disrupted river flows and altered hydrologic patterns. Most wetland losses in Louisiana have resulted from submergence, as accretion of new soil and organic plant material is unable to keep pace with the relative sea level rise because of altered hydrology, lack of mineral sediments, and deteriorated landscapes that do not support continued growth of marshes. More than 1,900 square miles of coastal land, mainly tidal wetlands, has been lost since the 1930s (Barras, 2006). The annual rate of loss slowed from a peak of 40 square miles per year in the 1960s and 1970s to 24 square miles per year between 1990 and 2000. In addition to these trends, the land and water configuration of coastal Louisiana was dramatically affected by Hurricanes Katrina and Rita in 2005. Comparison of satellite imagery before and after the landfalls of these hurricanes showed that the water area in coastal Louisiana increased by roughly 217 square miles after their passage (Barras, 2006).

Other Pollutants

In addition to concerns about nutrients and sediments, there are many other Mississippi River water quality problems. For example, toxic substances of major concern in the Mississippi River include metals (primarily mercury, zinc, and lead), organometallic compounds (primarily methylmercury and tributyltin), and a long list of toxic organic chemicals. Important among the latter are the chlorinated aromatic compounds (including PCBs), chlorinated hydrocarbons (including DDT, its degradation products, and other pesticides), and polycyclic aromatic hydrocarbons. Fecal bacteria also are an important water quality concern in areas of the mainstem Mississippi River. The distribution of all of these contaminants along the river depends on the nature and location of the source, their stability, their dilution by receiving waters, and their adsorption by sediments and the movement of these sediments.

The U.S. Geological Survey (USGS) Mid-Continent Survey of contaminants in the Mississippi River and some of its major tributaries was conducted from 1987 to 1990 and expanded in 1991-1992 to include samplings along the length of the river between Minneapolis-St. Paul and New Orleans (Meade, 1995), and provides a data-rich snapshot of conditions in the river. The survey focused on dissolved contaminants, those associated with the suspended sediments, and those stored in river bottom sediments in the upper Mississippi River. Much of the summary below is derived from this study and from the Meade (1995) synthesis volume.

Metals

Lead and other heavy metals are associated with suspended sediments along the length of the Mississippi River (Figure 2-14). Lead comes from both natural and human-related sources, but its sources in the upper Mississippi River are mostly industrial and municipal. Lead in suspended sediments tends to be most concentrated downstream of Minneapolis-St. Paul and also shows slight increases related to more concentrated inputs from the Ohio River. Lead at "moderately polluted" levels (40 micrograms per gram of sediment) occurs in bed sediments within the pools of the upper Mississippi River and is closely correlated with the finer clay fraction of sediments, but may also reflect the legacy of lead mining in some areas. Another metal of environmental concern, mercury, has been found at concentrations considerably lower than those of lead (Garbarino et al., 1995).

Dissolved inorganic mercury (Figure 2-15) was lowest in the Mississippi River's upper reaches and gradually increased downriver. High concentrations were measured downstream of tributaries, such as the Des Moines, Illinois, and Missouri Rivers, and near large metropolitan and industrial centers, specifically St. Louis, Vicksburg, and below Baton Rouge. Concentrations of dissolved inorganic mercury decreased below these points, due to transformation to organic forms, adsorption onto sediments, or both. Mercury concentrations in sediments of pools in the upper Mississippi River were correlated with the organic content of the sediments, and except in Lake Pepin, most were not high enough to cause adverse toxicological effects. Mercury can bioaccumulate in many aquatic organisms, especially fish, through ingestion of suspended or bed-sediment particles.

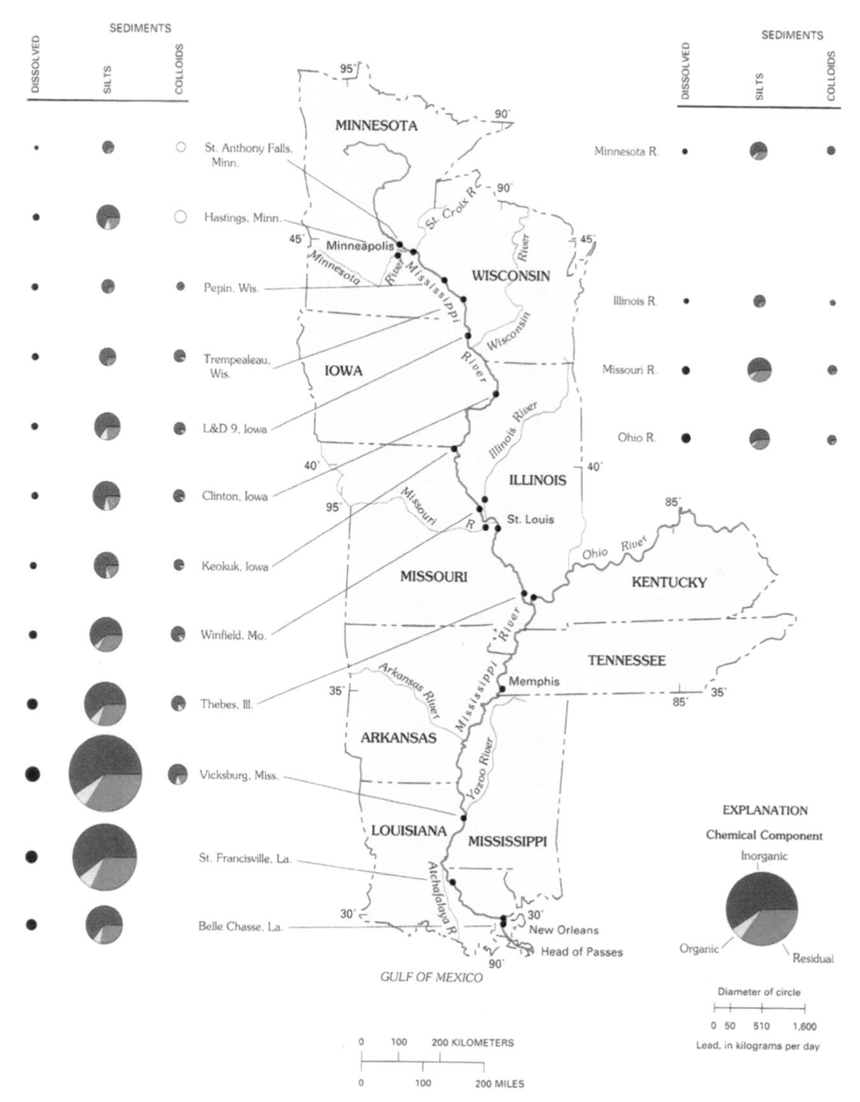

FIGURE 2-14 Lead in Mississippi River waters and sediments.
SOURCE: Garbarino et al. (1995).

PCBs

Polychlorinated biphenyls are organic contaminants that were formerly used widely in industrial applications. Along the Mississippi River, they are typically most highly concentrated in suspended sediments near Minneapolis and St. Louis. Industrial activities in the Minneapolis-St. Paul region

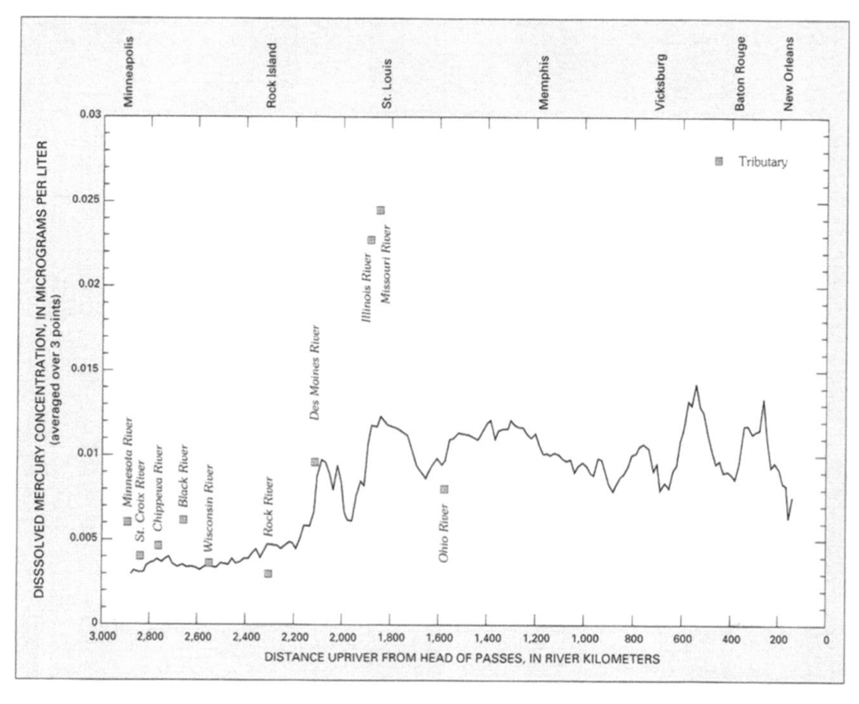

FIGURE 2-15 Mercury dissolved in Mississippi River water.
SOURCE: Garbarino et al. (1995).

led to PCB concentrations there that were five to ten times higher than in other parts of the river. Concentrations of PCBs were greatest in sediments between Minneapolis-St. Paul and Lake Pepin. Increased concentrations near St. Louis reflect the input of suspended sediments from the Ohio River, which usually contain more PCBs than do the waters in the middle reaches of the river. Hexachlorobenzene, another organic contaminant of industrial origin adsorbed to suspended sediments, is derived predominantly from the Ohio River and the industrial corridor along the lowermost 400 kilometers (248 miles) of the Mississippi River. There are hundreds of different kinds of PCBs, and numerous medical studies show that they have a variety of human health effects. In addition to the direct implications of PCBs for human health, bioaccumulation of PCBs in fish tissue is another key concern in the Mississippi River (see Box 2-1).

PCBs are legacy contaminants that are stored in bed sediments in the navigation pools of the upper Mississippi River. The concentrations in the upper (10 centimeters) sediments are high below Minneapolis-St. Paul, reach their highest values in Lake Pepin, and are significantly lower in the

BOX 2-1
Toxic Substances and Fish Contamination

A key concern for commercial and recreational fishermen on the Mississippi River is the existence of toxic substances in the river's fish populations. States along the Mississippi River issue various versions of fish consumption advisories, which are usually based on concentrations in fish tissue. Fish tend to accumulate long-lived, slightly soluble chemicals such as PCBs, pesticides, and herbicides in their fatty tissue. Concentrations of toxic substances in fish tissue can be much higher than in the water. Most of the 10 states along the mainstem Mississippi River list some reach as being of impaired water quality, and most of these impairments are based on fish tissues that contain unacceptable concentrations of toxic substances. For example, Illinois, Minnesota, Missouri, Tennessee, and Wisconsin list the entire river for PCBs; Tennessee lists dioxin and chlordane; and Minnesota and Wisconsin list mercury, all on the basis of high concentrations in fish.

pools downriver of Lake Pepin (Rostad et al., 1994). After they were banned in 1977, concentrations of PCBs in the upper layers of bed sediments decreased dramatically, especially in pools 2-9 (UMRCC, 2002). Evidence of the contaminant legacy, however, is seen in deeper buried sediments, where concentrations are much higher (Rostad et al., 1994). Chlordane concentrations also decreased, especially in pools 10-26 (UMRCC, 2002).

Pesticides and Herbicides

About two-thirds of all pesticides and herbicides used in U.S. agriculture, most of which are used for weed control, are applied in the Mississippi River basin (Goolsby and Pereira, 1995). Concentrations of 32 pesticides and herbicides and their degradation products have been found in Mississippi River water (Goolsby and Pereira, 1995); the most common is atrazine, a pre-emergent herbicide used mainly on cornfields. It is nearly ubiquitous along the river, with highest concentrations near St. Louis. It derives from the Missouri, Illinois, and other rivers that drain farming regions across the Corn Belt. Metolachlor, like atrazine, also was detected in more than 95 percent of the Goolsby and Pereira (1995) samples. Average annual concentrations of all pesticides and herbicides were far below the maximum contaminant levels for treated drinking water or health advisories, and only a few individual samples exceeded allowable levels of atrazine, alachlor, and cyanazine. Pesticide and herbicide concentrations are typically low during the summer, fall, and winter and then rise sharply in April and May as farmers apply them to fields for weed control and spring rains wash some of the chemicals off. Pesticide and herbicide concentrations then typically decline in June, depending on rainfall patterns. Unlike the legacy pollutants discussed earlier, most pesticides and herbicides in use today are water soluble and decay relatively rapidly.

Fecal Bacteria

Coliform bacteria are present in the fecal matter of all warm-blooded animals, including humans. Therefore, they are present in untreated or incompletely treated domestic sewage, animal waste (livestock, domestic and wild), and feedlot runoff. They have been used for nearly 100 years as an indicator of the possible presence of many pathogenic organisms that are too impractical to test for and quantify routinely. The only comprehensive collection of fecal coliform data for the entire Mississippi River is that compiled by the USGS for 1982-1992 (Barber et al., 1995; Figure 2-16). Those data indicated greatly improved water quality compared to levels measured in the preceding 80 years, although there were still high counts of fecal coliforms near and downstream of the Quad Cities (Bettendorf and Davenport, Iowa, and Moline and Rock Island, Illinois); below St. Louis and Cape Girardeau, Missouri; below Vicksburg, Mississippi; and below Baton Rouge and Belle Chasse, Louisiana.

In Minnesota, the Twin Cities Metropolitan Council has effected major improvements in Mississippi River water quality with improved wastewater treatment since the 1960s. Since then, fecal coliform counts at St. Paul gradually have trended downward. Water quality improvement at Newport-Inver Grove, Minnesota, downstream from the main wastewater treatment plant, has been even more dramatic. As a result of these improvements, Minnesota now lists only 36 miles of the Mississippi River as having impaired water quality because of fecal coliforms in the vicinity of the Twin Cities, all upstream of the main wastewater treatment plant. Further downstream in Illinois, several areas along the Mississippi River have fecal coliform counts with annual averages lower than the standard, but Illinois lists the entire river along its border as being of impaired quality due to fecal coliforms because of high counts during storm runoff. In the Mississippi River below Baton Rouge, Louisiana, geometric means at five stations were lower in 1984-1995 than in 1977-1984 (Caffey et al., 2002). An average of 200 to 500 fecal coliform colonies per 100 milliliters characterized the Mississippi River below Baton Rouge for 1982-1992 (Barber et al., 1995).

The fact that fecal coliform counts at many locations along the river routinely average more than 200 CFU (colony-forming units) per 100 milliliters does not necessarily mean that wastewater treatment plants are not effective enough. There is general agreement today that the major remain-

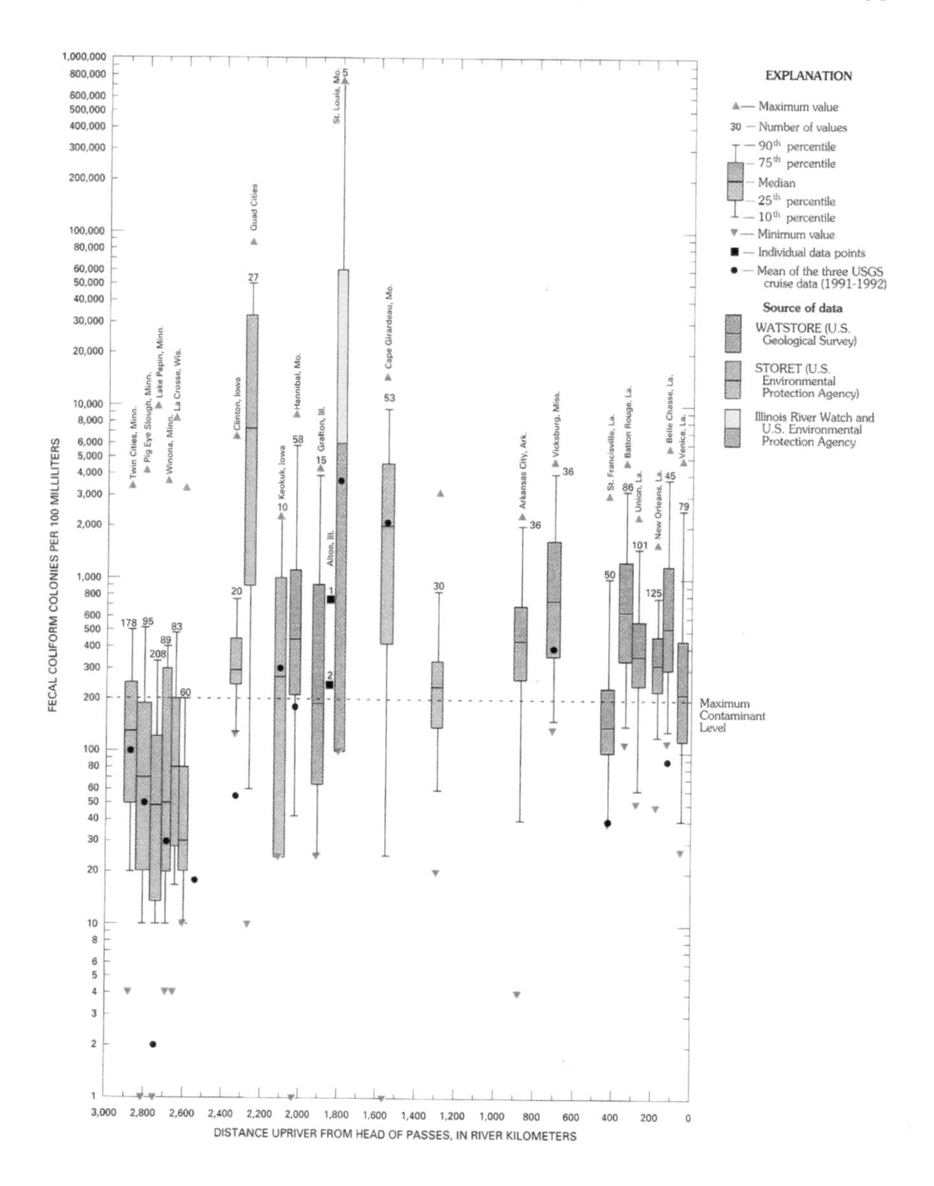

FIGURE 2-16 Fecal coliform concentrations along the Mississippi River from 1982 to 1992 (U.S. Environmental Protection Agency, STORET database; U.S. Geological Survey WATSTORE database; Illinois River Watch; specific samples from the 1991-1992 USGS study). The bar-and-whisker plots represent the median and 10th, 25th, 70th, and 90th percentiles.
SOURCE: Barber et al. (1995) (erratum resulted in this corrected Figure 53 from Barber et al., 1995).

ing fecal coliform sources derive from urban and rural stormwater runoff, followed by combined sewer overflows (CSOs) from some large cities, and separate sanitary sewer overflows (SSOs) in some cities, during major rainstorms. Sewer overflows are considered point sources under the Clean Water Act and are being addressed by many cities, but correction is slow and expensive. However, stormwater runoff is more difficult to control.

Emerging Contaminants

New types of chemical and biological contaminants are the subject of exploratory monitoring. Examples of emerging contaminants include pharmaceuticals, fluorochemicals, and human-animal antibiotics and hormones (Kolpin et al., 2002; Field et al., 2006). Such compounds have been measured in the Mississippi River and its tributaries (e.g., Boyd and Grimm, 2001; Kolpin et al., 2002). Potential concerns related to these entities include abnormal physiological processes and reproductive impairment, induction of cancer, development of antibiotic-resistant bacteria, and other effects. For many emerging contaminants, little is known about potential effects on humans and aquatic ecosystems, especially for long-term, low-level exposure, which is the typical scenario.

WATER QUALITY IMPACTS IN THE GULF OF MEXICO

The Mississippi River and its freshwater discharge, sediment delivery, and nutrient loads have strongly influenced the physical and biological processes in the adjacent Gulf of Mexico over geologic time and past centuries, and even more strongly during the last half of the twentieth century. As mentioned earlier, nutrient overenrichment in many areas around the world is having pervasive ecological effects on coastal ecosystems, including noxious (and possibly toxic) algal blooms, reduction in levels of dissolved oxygen, and subsequent impacts on living resources (NRC, 2000a; Vitousek et al., 1997). The largest zone of oxygen-depleted coastal waters in the United States, and the entire western Atlantic Ocean, is found in the northern Gulf of Mexico on the Louisiana-Texas continental shelf (Rabalais et al., 2002b; examples for 2001 and 2002 are shown in Figure 2-17).

The midsummer extent of bottom-water hypoxia (dissolved oxygen concentration less than 2 milligrams per liter) averages 12,900 square kilometers since systematic mapping began in 1985 and reached its maximal size to date of 22,000 square kilometers in 2002 (Rabalais and Turner, 2006; Figure 2-18). To appreciate the extent of these oxygen-depleted waters, consider that the size of this hypoxic zone is as large as New Jersey or Rhode Island and Connecticut combined and, at its largest, is the size of Massachusetts. The distance across the hypoxic area that stretches from the

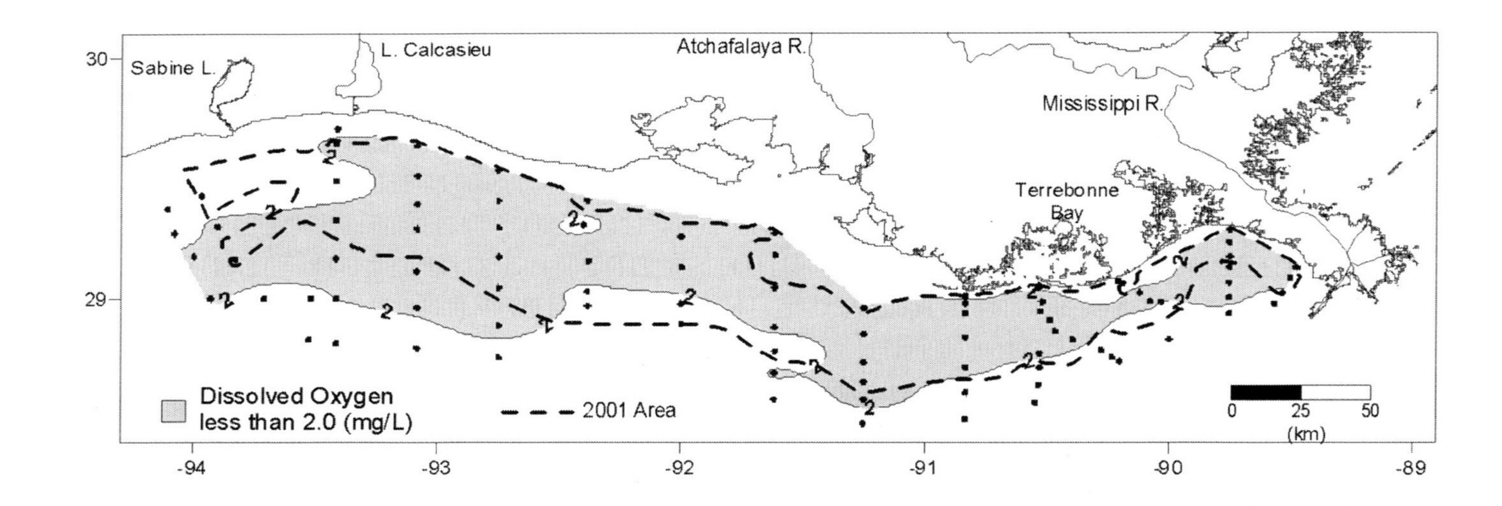

FIGURE 2-17 Similar size and expanse of bottom-water hypoxia in mid-July 2002 (shaded area) and in mid-July 2001 (outlined with dashed line).

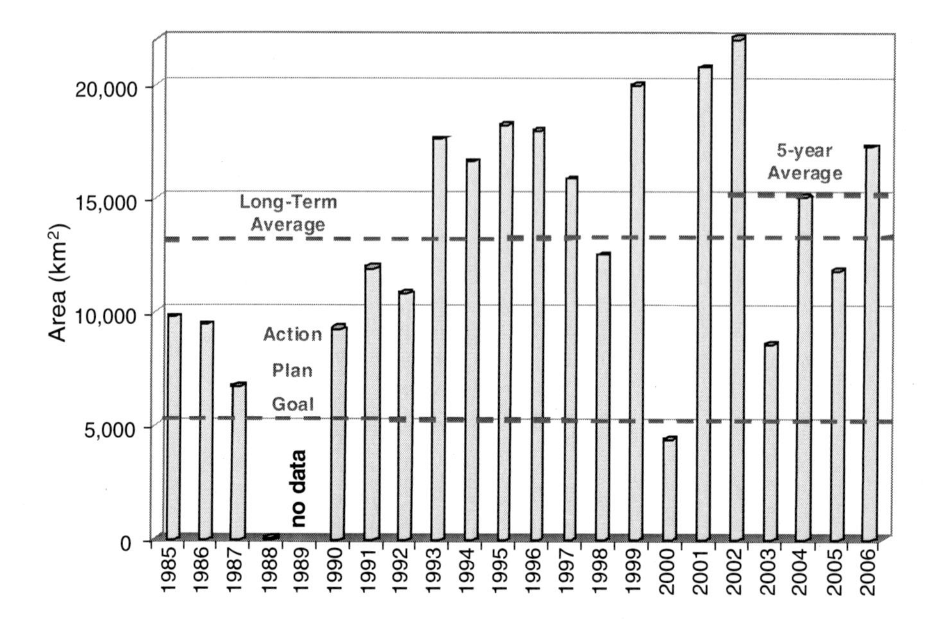

FIGURE 2-18 Estimated bottom areal extent of hypoxia (dissolved oxygen <2 mg/L) for midsummer cruises and the 2015 goal of 5,000 km^2 or less with long-term average sizes superimposed.
SOURCE: Modified, with permission, from Rabalais et al. (2002a). © 2002 by The American Institute of Biological Sciences.

Mississippi River across Louisiana's coast and onto the upper Texas coast is comparable to the distance between Chicago and St. Louis or between Milwaukee and Minneapolis-St. Paul.

The area affected by hypoxic, or low oxygen, conditions is commonly known as the Dead Zone because few marine animals can survive in these low oxygen concentrations (Rabalais and Turner, 2001). Swimming fish, crabs, and shrimp must escape or succumb to the low oxygen; other organisms eventually suffocate and die. The entire water column, however, is not devoid of oxygen, and fish survive in the upper waters along with hosts of bacteria at the seabed that can withstand low-oxygen conditions. Hypoxic conditions can damage fisheries and alter ecosystem functioning (Diaz and Rosenberg, 1995; Rabalais and Turner, 2001). Hypoxia, as a symptom of nutrient enrichment, is a growing problem around the world (Diaz and Rosenberg, 1995; Boesch, 2002; UNEP, 2006). The size and persistence of hypoxia on the Louisiana-Texas shelf, however, along with its connection to changes in Mississippi River nutrient delivery, make the Gulf of Mexico hypoxic zone a notable example.

Hypoxia is a seasonal but perennial feature of the coastal waters downstream from the Mississippi River discharge and is most prevalent from late spring through late summer. Typical water depths for hypoxia are between 5 and 40 meters. Although hypoxia is commonly perceived as a bottom-water condition, oxygen-depleted waters often extend up into the lower one-half to two-thirds of the water column. The effects, therefore, extend past organisms and processes at the bottom and into a much larger volume of water across the Louisiana coast.

The Mississippi River system is the dominant source of fresh water, sediments, and nutrients to the hypoxia zone in the northern Gulf of Mexico. The river carries 96 percent of annual freshwater discharge, 98.5 percent of total nitrogen, and 98 percent of total phosphorus load (calculated from U.S. Geological Survey streamflow data for 37 U.S. streams discharging into the Gulf of Mexico; Dunn, 1996; Rabalais et al., 2002b). Direct deposition of nitrogen from rainfall on the area of hypoxia is minimal (1 percent) compared to the load delivered by the Mississippi River (Goolsby et al., 1999). The river constituents are carried predominantly westward along the Louisiana-Texas coast, especially during peak spring discharge. Although the area of the discharge's influence is an open continental shelf, the magnitude of flow, ocean currents, and average 75-day residence time for fresh water result in an unbounded estuary, which is stratified for much of the year. This stratification is due primarily to salinity differences, and the stratification intensifies in summer with the warming of surface waters (Wiseman et al., 1997). Hypoxia is the result of the strong and persistent stratification coupled with the high phytoplankton growth in overlying surface waters that is fueled by river-derived nutrients (Rabalais and Turner, 2001; Rabalais et al., 2002a, 2002b). Nutrients delivered from the Mississippi River basin support phytoplankton growth in the immediate vicinity of the river discharges, as well as across the broader Louisiana and upper Texas coasts. The sinking of dead phytoplankton cells or the fecal pellets of zooplankton that have eaten phytoplankton to the lower water column and seabed provides a large carbon source for decomposition by oxygen-consuming bacteria. The bacterial decomposition process consumes dissolved oxygen in the water column at a higher rate than resupply from the upper water column across the stratified water layers. Oxygen levels slowly decline over days to weeks, eventually becoming less than the 2 milligrams per liter that defines hypoxia and may approach conditions without oxygen (anoxia).

The constituents of Mississippi River discharge changed substantially in the last half of the twentieth century, as outlined above. There is considerable evidence that nutrient-enhanced primary production, particularly by nitrate-nitrogen (nitrate-N), in the northern Gulf of Mexico is causally related to the oxygen depletion in the lower water column (CENR, 2000;

Justić et al., 2002; Rabalais et al., 2002a, 2002b; Turner et al., 2005, 2006). For example, strong temporal linkages have been demonstrated among freshwater delivery, nitrate flux, high algal production in the surface waters (Justić et al., 1993; Lohrenz et al., 1997), and subsequent bottom-water hypoxia (Justić et al., 1993). Models of a site within an area of persistent hypoxia about 100 kilometers west of the Mississippi River clearly link nitrate flux from the Mississippi River with both surface and bottom-water oxygen conditions (Justić et al., 1996, 2002). Other models have been used to predict oxygen conditions retroactively on the Louisiana coast to the early 1950s when nitrate data became readily available; all results show a decrease in bottom-water oxygen levels in the early 1970s (Scavia et al., 2003; Turner et al., 2005, 2006). These models effectively link nitrate loads from the Mississippi River with the bottom area size of the hypoxic zone in midsummer. Data showing oxygen concentrations on the Louisiana coast indicate a gradual decline in bottom-water oxygen levels across the coast for the periods of record (1982-2002 and 1978-1995; see Stow et al., 2005; Turner et al., 2005). A model developed by Turner et al. (2006) tests the relationship of hypoxic area size to factors such as other forms of nitrogen, phosphorus, dissolved silicate, and their concentration ratios. In this model, the strongest relationship was found with nitrate.

To understand conditions on the Louisiana coast for periods in which actual oxygen measurements do not exist, chemical and biological indicators in sediments where hypoxia is now a persistent condition were examined. The accumulated evidence in sediments shows trends of increased phytoplankton production in the last half of the twentieth century accompanied by more severe or persistent hypoxia beginning in the 1960s to 1970s and becoming most pronounced in the 1990s (Rabalais et al., 2007). The shifts in sediment indicators are temporally consistent with the rise in Mississippi River nitrate levels and with modeling results. Specific indicators demonstrate increased accumulation of phytoplankton biomass—stable carbon isotopes, silica, remains of diatoms, the abundance of a specific diatom that can generate harmful toxic substances, and specific phytoplankton pigments. These trends show that while there are signs of increased production and oxygen depletion earlier in the twentieth century, the most dramatic changes have occurred since the 1960s, when the nitrate concentration and load from the Mississippi River began to increase.

Hypoxia in the northern Gulf of Mexico occurs in an important commercial and recreational fisheries zone that accounts for 25 to 30 percent of the annual coastal fisheries landings for the United States. The ability of organisms to live, or even survive, either at the bottom or within the hypoxic water column is severely affected as the depletion of oxygen progresses toward anoxia. When the dissolved oxygen content is less than 2 milligrams per liter, animals capable of swimming evacuate the area. Less motile animals living in the sediments experience stress or die as oxygen concentrations fall to zero. The abundance of animals in the sediment and the diversity of the sediment-dwelling community are severely reduced, which means less food and less preferred food for the shrimp and fish that depend on them. Numerous studies document the effects of hypoxia on coastal fish and shrimp. Shrimp, as well as the dominant fish, the Atlantic croaker, are absent from the large areas affected by hypoxia (Renaud, 1986; Craig and Crowder, 2005; Craig et al., 2005). There is a negative relationship between the catch of brown shrimp—the largest economic fishery in the northern Gulf of Mexico—and the relative size of the midsummer hypoxic zone (Zimmerman and Nance, 2001). The catch per unit effort of brown shrimp declined during a recent interval in which hypoxia was known to expand (Downing et al., 1999). The presence of a large hypoxic water mass when juvenile brown shrimp are migrating from coastal marshes to offshore waters inhibits their growth to a larger size and thus affects the poundage of captured shrimp (Zimmerman and Nance, 2001). The unavailability of suitable habitat for shrimp and croaker forces them into the warmest waters inshore and also cooler waters offshore of the hypoxic zone with potential effects on growth, trophic interactions, and reproductive capacity (Craig and Crowder, 2005). The overall implications of these indirect stressors for the Gulf of Mexico fisheries production and its overall productivity are not fully known. There have been no catastrophic losses of fishery resources in the northern Gulf of Mexico. In fact, the abundance of some pelagic components, which have greater volume but less economic value, has increased. This has been to the detriment of bottom-dwelling animals (Chesney and Baltz, 2001).

Several different initiatives have been taken to help address the problem of hypoxia on the Louisiana coastal shelf. For example, the Action Plan for Reducing, Mitigating, and Controlling Hypoxia in the Northern Gulf of Mexico (USEPA, 2001) was endorsed by federal agencies, states, and tribal governments. The action plan calls for a long-term adaptive management strategy that couples management actions with enhanced monitoring, modeling, and research. Implementation will depend on a series of voluntary and incentive-based activities, designed within a series of subbasin strategies, including best management practices on agricultural lands, wetland restoration and creation, river hydrology remediation and riparian buffer strips, and stormwater and wastewater nutrient removal (Mitsch et al., 2001). These subbasin efforts, which are intended to achieve a nitrogen load reduction of 30 percent, will work toward a goal of a Gulf of Mexico hypoxic zone smaller than 5,000 square kilometers (five-year running average) by the year 2015. Some modeling studies, however, suggest that a greater reduction—35 to 45 percent—in the nitrogen load will be required to meet this goal (Justić et al., 2003; Scavia et al., 2003). In 2006, five years after its adoption, the action plan was being reassessed with regard to new scientific knowledge and management scenarios. Despite the plan and the activities begun in connection with it, in the last five years little change has been implemented within the watershed, and the size and persistence of the hypoxic area continue unabated.

SUMMARY

The Mississippi River basin covers nearly one-half of the continental United States and exhibits a variety of landforms, landscapes, climate zones, and land uses. There are natural differences in these features across the watershed, and there have been extensive human-induced changes in land uses and Mississippi River hydrology. Huge swaths of forested lands and prairie have been converted to cropland; numerous locks and dams have been constructed on the upper Mississippi, Missouri, and Illinois Rivers; most of the natural wetlands along the length of the river and in the watershed have been drained and converted to other uses; and huge levees for both flood protection and navigation purposes have been constructed along the lower Mississippi River. The primary land use across the basin today is agriculture. With regard to human population, many parts of the Mississippi River basin are lightly populated in comparison with the more urbanized U.S. East and West Coasts, and urban areas constitute only a small percentage of total land use in the basin. Population in all the basin states is growing; while some rural areas are experiencing population declines, some urban areas are growing rapidly. Differences in natural features across the river basin, coupled with two centuries of anthropogenic changes in land cover, land uses, and the construction of river control structures, influence both the amount of Mississippi River discharge and its constituents and pollutants, such as nutrients, suspended sediments and other particulate materials, and toxic chemicals.

In terms of Mississippi River hydrology and sediment transport, the river exhibits a very different character in its various reaches. The upper and lower Mississippi Rivers are, in fact, in many ways two different river systems. For example, many portions of the upper Mississippi River contain islands and large backwater areas important to recreational activities such as boating, fishing, and trapping, and they share the river, its channel, and its numerous navigation pools with commercial navigation. By contrast, the lower Mississippi River below Cairo, Illinois, contains fewer islands and is leveed off from most of its previous floodplain areas. The lower Mississippi River carries much larger river flows and poses dangers that inhibit recreational boating, fishing, and related activities.

Levels of sediment transported by the Mississippi River and its tributaries have changed greatly since the 1700s. In particular, whereas the Missouri River once delivered huge quantities of sediment into the Mississippi River, construction of storage dams on the Missouri River in the 1950s and 1960s greatly reduced these inputs. The total amount of sediment carried by the Mississippi River and delivered to the Gulf of Mexico has been reduced significantly. The depletion of this sediment, among other natural and human activities, has led to the loss of many wetlands and coastal barriers in coastal Louisiana and other areas along the U.S. Gulf Coast. The upper Mississippi River today carries a proportionally greater amount of the river's total sediment load than in 1700, and sedimentation is a problem in many areas of the upper Mississippi River, both in the main channel and in backwater areas.

Highest inputs and concentrations of nutrients are in the upper and middle reaches of the Mississippi River. Uptake and transformation of nutrients is more likely to occur closer to the sources and in the smaller streams. Once nutrients reach the mainstem, there is little loss or dilution on the way to the river delta—an important point to be considered in nutrient management efforts.

Excess nutrient input to the Mississippi River, in various forms of dissolved and particulate nitrogen and phosphorus, causes significant water quality problems both within the Mississippi River itself and in the coastal waters of the northern Gulf of Mexico. These latter problems manifest themselves as Gulf of Mexico hypoxia, one of the nation's prominent regional-scale water quality problems. Nutrient enrichment, primarily from dissolved inorganic nitrogen, causes disturbance of the coastal ecosystem including, but not limited to, hypoxia, noxious and toxic algal blooms, impacts on living resources, and fishery impacts. The importance of phosphorus as a limiting nutrient to phytoplankton growth is more evident in the spring and in the upper Mississippi River. Given the importance of both nitrogen and phosphorus in various forms, it is necessary to consider management of both of these nutrient inputs, which stem primarily from nonpoint sources.

These activities and modifications contribute to water quality problems along the river's mainstem that are numerous, variable in nature, and of different magnitudes in different parts of the river. These problems can be divided into three broad categories: (1) contaminants with increasing inputs along the river that accumulate and increase in concentration downriver from their sources (e.g., nutrients and some fertilizers and pesticides); (2) legacy contaminants stored in the riverine system, including contaminants adsorbed onto sediment and stored in fish tissue (e.g., PCBs and DDT); and (3) "intermittent" water constituents that can be considered contaminants or not, depending on where they are found in the system, at what levels they exist, and whether they are transporting adsorbed materials that are contaminants. The most prominent component in the latter category is sediment. In some portions of the river system, sediment is overly abundant and for that reason can be considered a contaminant. In other places it is considered a natural resource in deficient supply.

At the scale of the entire Mississippi River, including its effects that extend into the northern Gulf of Mexico, nutrients and sediment are the two primary water quality problems. Nutrients are causing significant water quality problems within the Mississippi River itself and in the northern Gulf of Mexico. With regard to sediment, many areas of the upper Mississippi River main channel and its backwaters are experiencing excess sediment loads and deposition, while limited sediment replenishment is a crucial problem along the lower Mississippi River and into the northern Gulf of Mexico. Nutrients and sediments from nonpoint sources are the primary water quality problems focused on in this report. With respect to nutrients and sediments (and some toxic substances), water quality in the lower Mississippi River is determined largely by inputs in the upper Mississippi River basin, with different portions of the upper river basin having a dominant influence for particular constituents. For example, sediment loads are determined largely by the Missouri River contributions, and nutrient contributions are primarily from the upper Mississippi River.

In addition to nutrient and sediment issues, the Mississippi River has a variety of other water quality challenges. Toxic substances, including PCBs, metals, and pesticides, have important human health implications and are related primarily to legacy inputs. Their concentrations, fortunately, have been decreasing with time, in large part due to reductions in point source contributions as a result of the Clean Water Act. Similarly, the Clean Water Act has been useful in substantially reducing fecal coliform levels in the Mississippi River. The Clean Water Act was designed to remediate some of the impacts of human activities and has been effective in reducing many impacts attributable to point sources. Many of today's water quality problems, however, are nonpoint in nature.

Whereas the Clean Water Act has been successful in reducing many point source pollution problems along the Mississippi River, it has not been as successful in reducing nonpoint source pollutants. Both the source and the scale of Mississippi River and Gulf of Mexico nonpoint source water quality problems pose significant Clean Water Act-related management challenges. The following chapters describe the Clean Water Act and discuss challenges in its administration to achieve its goals of attaining fishable and swimmable water quality and restoring the chemical, physical, and biological integrity of water resources as these goals apply to the Mississippi River.

3

The Clean Water Act

Congress first enacted the Federal Water Pollution Control Act (FWPCA) in 1948. Substantial amendments to that act—first in 1972 and again in 1977—created the statute now known as the Clean Water Act. Congress amended the FWPCA repeatedly from 1956 on; however, substantial amendments in 1972 created the contemporary structure of the act, which acquired the name "Clean Water Act" in 1977. The 1972 amendments represented a significant change in approach in that they shifted the emphasis in water quality regulation from an earlier focus on state-level water quality standards, to a federal permitting scheme setting technology-based and water quality-based effluent limits for individual dischargers. Moreover, Congress designed the 1972 act "to restore and maintain the chemical, physical, and biological integrity of the Nation's waters" (Section 101(a)). The Clean Water Act authorizes water quality programs, requires state water quality standards, requires permits for discharges of pollutants into navigable waters, and authorizes funding for wastewater treatment works, construction grants, and state revolving loan programs. The act underwent subsequent amendments in 1981, 1987, and 1990.

The U.S. Environmental Protection Agency (EPA), the U.S. Army Corps of Engineers, and the states are jointly responsible for implementing the Clean Water Act and for achieving the goals of attaining water quality that is, according to the act, at least "fishable and swimmable." In general, the Clean Water Act requires states to establish water quality standards for waters within their borders by designating specific uses for their waters (so-called designated uses) and establishing criteria by which to protect those uses, control pollutant sources, and monitor and assess water quality.

States are responsible for submitting periodic water quality assessment reports—Section 305(b) reports—and lists—Section 303(d) lists—of impaired waters to the EPA. They then are supposed to restore impaired waters by developing Total Maximum Daily Loads (TMDLs), which are limits that in theory, if fully implemented, should ensure that the state's waters achieve the relevant quality standards. The EPA establishes federal guidance water quality criteria and oversees the establishment of state water quality standards to ensure that they are consistent with the requirements of the Clean Water Act, including ensuring that state-adopted water quality criteria are sufficient to attain the designated uses assigned by the state.The EPA also oversees state National Pollutant Discharge Elimination System (NPDES) permitting, issuing NPDES permits to dischargers in states that have not assumed this permitting authority and helping to resolve interstate water pollution issues. Finally, the U.S. Army Corps of Engineers implements the "dredge-and-fill" (wetlands) permit program in almost all states, subject to EPA oversight.

The Clean Water Act (CWA) is a lengthy and complex body of legislation, and this chapter does not attempt to examine all of its provisions. Instead, for purposes of this report and its emphasis on the Mississippi River, the chapter focuses on the CWA sections and the federal and state authorities and responsibilities that are important in understanding Clean Water Act applications and challenges along the Mississippi River. This report focuses on point and nonpoint source pollution of the mainstem Mississippi River, not ancillary issues that may arise with regard to the dredging and filling of wetlands. As a result, at the federal level, this report focuses on EPA's regulatory authority, not that of the Corps of Engineers. The EPA's jurisdiction to regulate discharges of pollutants into the Mississippi River and its major tributaries is clear, despite recent U.S. Supreme Court decisions and agency guidance regarding the extent of federal jurisdiction over wetlands and isolated waters. This chapter also discusses interstate and federal-state water quality interactions and the relevance of the CWA to these interactions. The chapter is divided into four sections: origins of the Clean Water Act; Federal Water Pollution Control Act amendments of 1972; state-level authority in protecting water quality; and interstate water quality protection.

ORIGINS OF THE CLEAN WATER ACT

The Refuse Act

Congress enacted the Rivers and Harbors Act of 1899 to preserve and enhance navigation in the nation's waters. Section 13, the Refuse Act, prohibits pollution of the nation's "navigable waters." The language of Section

13 is broad, and throughout the 1960s the federal government increasingly used it to prosecute water pollution cases. In an attempt to formalize the federal government's use of the Refuse Act to address water pollution, President Richard Nixon in 1970 ordered the Corps of Engineers and the administrator of the newly formed Environmental Protection Agency "to implement a permit program . . . to regulate the discharge of pollutants and other refuse matter into the navigable waters of the United States or their tributaries" (Nixon, 1970). These agencies promulgated their regulations within a year, creating the first federal water pollution permit program in the United States.

Nevertheless, despite the breadth of the Refuse Act's language and Supreme Court rulings upholding the use of that act to punish polluters, the Rivers and Harbors Act's focus on navigation limited its usefulness for water quality regulation. In particular, the only waters subject to the Refuse Act are waters that are or can be made navigable-in-fact, including waters subject to the ebb and flow of the tide (33 C.F.R. Section 329.4). Thus, the Refuse Act could not address water quality problems comprehensively, even with the new permit program. This prompted Congress to expand the definition of regulated "navigable waters" in the Clean Water Act to encompass "the waters of the United States, including the territorial seas" and at least some non-navigable-in-fact waters, as discussed more fully below.

The Federal Water Pollution Control Act of 1948

Congress addressed more general water quality concerns through the Federal Water Pollution Control Act of 1948. The FWPCA was not a regulatory program, however; instead, its primary purpose was to encourage *states* to improve water quality, largely through federal grants and loans for the construction of publicly owned treatment works (POTWs, or sewage treatment plants; FWPCA, 1948). Under this act, the federal Surgeon General (the EPA did not exist until 1970) *could* institute abatement actions, but only to protect interstate waters and only to abate pollution "which endangers the health or welfare of persons in a state other than that in which the discharge originates" (FWPCA, 1948, Section 2(d)).

Congress amended the FWPCA in 1952, 1956, 1961, 1965, 1966, and 1970, slowly expanding the federal government's abatement authority. In 1961, for example, Congress allowed the Secretary of Health, Education, and Welfare to bring abatement actions when pollution of *any* navigable-in-fact water (as opposed to interstate waters) affected the health or welfare of any person (FWPCA Amendments, 1961). In 1966, federal enforcement authority expanded again; this time Congress gave the Secretary of the Interior authority to take abatement actions to control pollution of international waters, through the Clean Water Restoration Act of 1966 (CWRA, 1966).

However, until 1972 Congress had never created a general federal permit program to control water pollution. Instead, in 1965 Congress amended the FWPCA to create a state-focused, water quality standards approach to water quality regulation (WQA, 1965). Under these amendments, states could continue to receive federal grants and loans to aid in water quality improvements, but now only if they established water quality standards. However, the federal government could eventually set water quality standards for any states that refused to do so. Slow progress by the states in establishing water quality criteria and related programs raised interest in a technology-based regulatory approach and prompted the passage of amendments in 1972. Another significant problem with the 1965 water quality standards program was the difficulty of enforcing an ambient standard regime without source-specific limitations.

FEDERAL WATER POLLUTION CONTROL ACT AMENDMENTS OF 1972

Congress enacted the contemporary version of the Clean Water Act through the Federal Water Pollution Control Act Amendments of 1972 (FWPCA, 1972), which set out "to restore and maintain the chemical, physical, and biological integrity of the Nation's waters" (Section 101(a)). More specifically, the act established "national goal[s] that the discharge of pollutants into the navigable waters be eliminated by 1985" and "that wherever attainable, an interim goal of water quality which provides for the protection and propagation of fish, shellfish, and wildlife and provides for recreation in and on the water be achieved by July 1, 1983"—the act's so-called (and still unattained) fishable and swimmable goal (Section 101(a)(1), (2)). The 1972 amendments pursued these goals by transforming the FWPCA's previously state-focused water quality standards approach into a federal permitting scheme based primarily on end-of-the-pipe effluent limitations for individual dischargers (Craig, 2004).

Table 3-1 lists the major sections of the Clean Water Act. This table illustrates clearly the act's broad scope. It encompasses sewage and industrial waste treatment (Title II), point source discharge permitting (Section 402), ambient water quality objectives (Section 303), state water quality standards, TMDLs and reporting requirements (Sections 303 and 305), nonpoint source management (Section 319), water quality in estuaries (Section 320), ocean discharges (Section 403), wetland protection (Section 404), and other aspects of protection and restoration of surface water quality in the United States.

TABLE 3-1 Major Clean Water Act Provisions

Section 101, 33 U.S.C. § 1251	Congressional goals and policies
Section 103, 33 U.S.C. § 1253	Interstate cooperation
Section 106, 33 U.S.C. § 1256	Grants for pollution control programs
Title II, §§ 201-221, 33 U.S.C. §§ 1281-1301	Grants for the construction of POTWs
Section 208, 33 U.S.C. § 1288	Areawide waste treatment management programs
Section 301, 33 U.S.C. § 1311	Act's basic prohibition; technology-based effluent limitations
Section 302, 33 U.S.C. § 1312	Water quality-based effluent limitations
Section 303, 33 U.S.C. § 1313	Water quality standards; TMDLs
Section 305, 33 U.S.C. § 1315	State water quality reporting
Section 309, 33 U.S.C. § 1319	Enforcement
Section 319, 33 U.S.C. § 1329	Nonpoint source management programs
Section 320, 33 U.S.C. § 1330	National estuary program
Section 401, 33 U.S.C. § 1341	State certification of federally authorized activities
Section 402, 33 U.S.C. § 1342	NPDES permit program
Section 403, 33 U.S.C. § 1343	Ocean discharge criteria
Section 404, 33 U.S.C. § 1344	Dredge-and-fill permit program
Section 502, 33 U.S.C. § 1362	Definitions
Section 505, 33 U.S.C. § 1365	Citizen suits
Section 510, 33 U.S.C. § 1370	State authority

Three aspects of the current Clean Water Act are particularly relevant to Mississippi River water quality and are discussed in the following sections: (1) the sewage treatment works or POTW construction grant and loan programs (which carried over from the pre-1972 versions of the FWPCA); (2) the two federal permit programs incorporated in the Clean Water Act, especially the NPDES permit program, but also the Section 404 dredge-and-fill permit program; and (3) a continuing water quality standards program, with new provisions to ensure linkage between permitting and overall water quality protection.

Federal Funding for Sewage Treatment Plants

The National and Interstate Problem of Sewage Pollution

The direct discharge of untreated sewage into the nation's waterways was a well-recognized public health problem that stimulated water quality protection legislation long before Congress enacted even the original FWPCA in 1948. Indeed, uncontrolled and lightly controlled discharges of sewage into interstate waterways—especially the Mississippi River and its tributaries—inspired much of the nation's pre-1972 interstate pollution law, based on the federal common law of nuisance. Progress in addressing the problem of untreated sewage discharges was steady but slow prior to the Clean Water Act of 1972.

Fittingly, it was an interstate sewage pollution case that established the Clean Water Act's supremacy regarding the regulation of water pollution. In 1972, Illinois sued four Wisconsin cities and two Wisconsin sewage commissions regarding their sewage pollution of Lake Michigan. By the time the case made its way to the Supreme Court for a final resolution almost a decade later, the Clean Water Act was firmly established as Congress's comprehensive regulatory regime for protecting and restoring water quality. Thus, while interstate water pollution lawsuits involving sewage pollution remain an important aspect of interstate water quality interactions, the Clean Water Act's requirements for sewage treatment and for interstate negotiations now control such conflicts (Craig, 2004).

The Clean Water Act and Sewage Treatment

Congress's approach to addressing sewage pollution through the Clean Water Act has been twofold. First, because collection and treatment of sewage generally is considered a government responsibility, Congress has provided funding to state and local governments to improve their sewage treatment capacity through the construction and improvement of POTWs. Second, Congress subjected these POTWs to a number of regulatory requirements to reduce the impact of their discharges on the nation's waterways.

With respect to sewage treatment capacity, the federal government began funding the construction of POTWs as early as the 1956 amendments to the FWPCA (WPCAA, 1956). However, Congress greatly expanded this grant program (known as the "construction grants program") in Title II of the Clean Water Act of 1972. Although grants initially were available for any POTW-related project, from October 1984 on, grants "shall be made only for projects for secondary treatment or more stringent treatment, or any cost effective alternative thereto, new interceptors and appurtenances, and infiltration-in-flow correction" (Section 201(g)). Title II authorized grant funds through FY 1990, ranging from $1 billion to $7 billion per year, which could pay up to 55 percent of each project's total costs (Section 207). By 1999, Congress had authorized $65 billion and had appropriated $73 billion for the grant program, resulting in the construction of thousands of new POTWs.

However, by 1989, the Title VI Clean Water State Revolving Fund (CWSRF) program had replaced the Title II grants program. Under this program, the EPA administrator "make[s] capitalization grants to each State" (Section 601(a)), which the states can then use to make loans to municipalities for three purposes: (1) to construct or improve POTWs; (2)

to implement states' nonpoint source management program; and (3) to develop management plans under the National Estuary Program. To receive the initial grant, each state had to agree to a number of conditions, including matching at least 20 percent of the grant with state funds.

Congress initially authorized a total of $8.4 billion for CWSRF capitalization grants for FY 1989 through FY 1994 (Section 607), but it appropriated far more. "Since 1987, states have used 96 percent (about $50 billion) of their CWSRF dollars to build, upgrade, or enlarge conventional wastewater treatment facilities and conveyances. Projects to build or improve wastewater treatment plants alone account for over 60 percent of this amount, with the remainder supporting the construction or rehabilitation of sewer and storm water collection systems" (GAO, 2006a). States use the remaining 4 percent for nonpoint source control in Section 319 programs. The Title II grant program and the Title VI CWSRF program have financed the construction and improvement of thousands of POTWs. For example, in 1972, only 32 percent of the nation's population was served by sewage treatment plants; by 1998, 74 percent of the population had such service (USEPA, 2003a). This program thus produced measurable improvements in the quality of the nation's waters, including the Mississippi River (Meade, 1995). The program—specifically, the construction and improvement of sewage treatment works—also resulted in some improvements to fish populations in the Mississippi River basin (see, for example, Lerczak and Sparks, 1995).

Beyond the construction of sewage treatment infrastructure is the issue of adequate sewage treatment within existing POTWs. Such treatment can be complicated by combined sewer overflow (CSO) events, which occur because many older sewer systems carry both sewage and stormwater runoff to the POTW. Although these systems normally are designed to handle small storm events, large storms often result in untreated discharges to surface waters. Moreover, in addition to the domestic sewage, POTWs receive industrial wastes, including some toxic pollutants that are discharged directly to waterways during CSO events. POTWs also can channel toxic pollutants into waterways from indirect dischargers. The Clean Water Act addresses these POTW-related water quality problems in three main ways.

First, POTWs that discharge into the nation's waters are subject to the act's National Pollutant Discharge Elimination System permit program, described below. Moreover, NPDES permits for POTWs must contain effluent limitations at least as stringent as secondary treatment (Section 301(b)(1)), requiring POTWs to engage in biological treatment of the sewage in addition to settling of particles in primary treatment.

Second, Congress amended the Clean Water Act specifically to address CSO problems. The act now requires that "each permit, order, or decree issued pursuant to [the Clean Water Act] . . . for a discharge from a municipal combined storm and sanitary sewer shall conform to the [EPA's 1994] Combined Sewer Overflow Control Policy" (Section 402(q)). The EPA's policy required that POTWs establish nine minimum controls on combined sewer systems by January 1, 1997 (USEPA, 1994). In addition to the nine minimum controls, communities with combined sewer systems are required to develop long-term plans to control combined sewer overflow events as necessary to meet water quality standards (USEPA, 1994). The control of combined sewer overflows, however, remains a significant national water quality issue (USEPA, 2003a).

Third, indirect dischargers—industrial dischargers that discharge into sewers leading to POTWs rather than into waterways—have been recognized as a threat to water quality almost as long as the 1972 Clean Water Act has been in existence. Indirect dischargers must pretreat effluent before sending it to the POTW in order to eliminate or reduce pollutants—generally toxic pollutants—"which are determined not to be susceptible to treatment by such treatment works or which would interfere with the operation of such treatment works" (Section 307(b)). As such, the pretreatment program seeks to eliminate "pass-through" pollution problems that otherwise occur when industrial dischargers, often seeking to avoid having to obtain their own NPDES permit, interfere with or pass through the POTWs.

Federal Permit Programs for Point Sources

The Clean Water Act's most basic prohibition for individual dischargers states that "except as in compliance with [the Act], the discharge of any pollutant by any person shall be unlawful" (Section 301(a)). Behind this seemingly simple prohibition, however, are several definitional complexities.

The Clean Water Act defines a "person" to be "an individual, corporation, partnership, association, state, municipality, commission, political subdivision of a State, or any interstate body" (Section 502(5)). Notably absent from this list is the federal government, but the Clean Water Act does require federal facilities to comply with the act's requirements "in the same manner, and to the same extent as any nongovernmental entity" (Section 313). Thus, persons covered by the Clean Water Act are broadly defined. More importantly, the Clean Water Act (Section 502(12)) defines the phrase "discharge of a pollutant" to mean:

> (A) any addition of any pollutant to navigable waters from any point source, (B) any addition of any pollutant to the waters of the contiguous zone or the ocean from any point source other than a vessel or other floating craft.

The Clean Water Act further defines most of the terms in this definition.

First, "point source" is defined broadly to include "any discernible, confined, and discrete conveyance" (Section 502(14)) such as pipes and ditches. Although this definition is broad, it does not cover all sources of water pollutants. By negative implication, any source of water pollutants that is not a point source is a nonpoint source, and the Clean Water Act's focus on "discernible, confined, and discrete conveyances" generally means that diffuse sources of water pollutants, such as agricultural or urban runoff or atmospheric deposition, do *not* qualify as point sources. The Clean Water Act assigns regulation of nonpoint source pollution to the states (Section 319).

Second, like point source, the Clean Water Act defines "pollutant" broadly, to include "dredged spoil, solid waste, incinerator residue, sewage, garbage, sewage sludge, munitions, chemical wastes, biological materials, radioactive materials, heat, wrecked or discarded equipment, rock, sand, cellar dirt and industrial, municipal, and agricultural waste discharged into water" (Section 502(6)). Given the breadth of the definitions of person, point source, and pollutant, the Clean Water Act's basic prohibition effectively prohibits all human-controlled additions of almost any material into the navigable waters, the contiguous zone, and the ocean, with limited exceptions.

Third, Congress purposely expanded the Clean Water Act's jurisdictional waters beyond those included in the Rivers and Harbors Act of 1899. Specifically, "navigable waters" for the Clean Water Act are "the waters of the United States, including the territorial seas" (Section 502(7)). In turn, the "territorial seas" are the first 3 miles of ocean extending from shore (Section 502(8)). As a practical matter, the Clean Water Act's navigable waters designate all of the waters that generally are subject to state jurisdiction, including both the inland waters (lakes, rivers, streams, and some wetlands) and, at least roughly, the offshore coastal waters whose submerged lands were given to states by Congress in the Submerged Lands Act (SLA, 2006).

The definition of navigable waters has become controversial as it applies to intrastate and apparently isolated wetlands or other waterbodies from both a statutory and a constitutional point of view. Since the 1985 decision in *United States v. Riverside Bayview Homes* (474 U.S. 121), the U.S. Supreme Court has struggled with the issue of how far CWA jurisdiction extends over wetlands and waterbodies that are obviously not navigable-in-fact. In that unanimous decision, the court held that the Clean Water Act extends to wetlands adjacent to navigable waters. In contrast, in *Solid Waste Agency of Northern Cook County v. U.S. Army Corps of Engineers* (531 U.S. 159 (2001)), a bare majority of justices decided that the Clean Water Act does *not* extend to isolated waters—specifically, to ponds that were not hydrologically connected to navigable waters. Most recently, in *Rapanos v. United States* (04-1034 U.S. 04-1384, 126 S. Ct. 2208 (2006)), the justices split 4-1-4 regarding the Clean Water Act's applicability to wetlands adjacent to nonnavigable *tributaries* of traditionally navigable waters, leaving lower courts and regulators with no clear test for the act's statutory and constitutional limitations. The proper constitutional balance between the states and the federal government—federalism—clearly was of concern.

With respect to navigable, interstate rivers such as the Mississippi, however, Congress's constitutional authority to regulate to protect water quality is uncontested. Moreover, the connection of upstream waters to the Mississippi River has been used to justify CWA jurisdiction over many tributaries and, more controversially, upstream wetlands. Indeed, in the 2006 *Rapanos* Supreme Court decision, concurring Justice Kennedy argued for precisely this approach, noting in particular the importance of wetlands to Mississippi River and Gulf of Mexico water quality issues. "Important public interests are served by the Clean Water Act in general and by the protection of wetlands in particular. To give just one example, ". . . nutrient-rich runoff from the Mississippi River has created a hypoxic, or oxygen-depleted, 'dead zone' in the Gulf of Mexico that at times approaches the size of Massachusetts and New Jersey. . . . Scientific evidence indicates that wetlands play a critical role in controlling and filtering runoff" (126 S. Ct. at 2246-47 [citations omitted]).

The other two types of waters that the Clean Water Act covers are the "contiguous zone" and the "ocean." Congress defined contiguous zone through a reference to the 1958 United Nations Convention on the Law of the Sea (Section 502(9)), and this zone refers to the area of ocean beyond the territorial sea, out to 12 miles. The ocean, in turn, is "any portion of the high seas beyond the contiguous zone" (Section 502(10)). In concert with what it claims to be customary international law, the United States asserts jurisdiction over a 200-nautical-mile-wide exclusive economic zone (EEZ) (Reagan, 1983) and has claimed a 200-nautical-mile-wide exclusive fishing zone since at least 1976 (see MSA, 1976). Thus, the Clean Water Act's regulatory program extends 200 nautical miles out to sea (Craig, 2004).

With respect to the Mississippi River, and in the context of this report, the Clean Water Act's marine coverage is most relevant to the Mississippi River's effects on the Gulf of Mexico. The act's extension to the oceans gives the federal government legal authority to regulate water quality in the Gulf of Mexico out to 200 nautical miles. Moreover, state regulatory authority under the Clean Water Act extends only through the territorial sea, or 3 nautical miles offshore, although Florida and Texas do have more far-reaching state jurisdiction over the waters of the Gulf of Mexico, extending to 3 marine leagues or about 12 nautical miles. Thus, consideration of Gulf of Mexico water quality generally necessitates state and federal cooperation, although Gulf of Mexico hypoxia occurs primarily in federal waters.

Section 404 Dredge-and-Fill Permit Program

The Clean Water Act's more limited permit program is the Section 404 permit program, more colloquially referred to as the dredge-and-fill or wetlands permit program. Under this program, the Secretary of the Army, acting through the U.S. Army Corps of Engineers, has the authority to "issue permits . . . for the discharge of dredged or fill material into the navigable waters at specified disposal sites" (Sections 404(a), 404(d)). States may acquire limited dredge-and-fill permitting authority (Section 404(g)), but to date only two states—Michigan and New Jersey—have done so (USEPA, 2007a). As a result, Section 404 permits remain almost entirely *federal* permits.

The U.S. EPA oversees the Section 404 permitting program in two ways. First, it issued guidelines that govern all Section 404 permitting (Section 404(b)(1)), and these guidelines require dischargers of dredged and fill material to minimize their adverse impacts on aquatic ecosystems (40 C.F.R. Section 230.1(c)). Second, the EPA has the authority to veto any particular Section 404 permit for a proposed discharge (Section 404(c)), although it has exercised this authority only rarely.

Given the CWA definition of "navigable waters," the Section 404 permit program does not apply more than 3 nautical miles out to sea (Section 502(7); 33 C.F.R. Section 328.4(a)). In addition, the Clean Water Act also limits this permit program to discharges of dredged or fill material, eliminating discharges of all other Clean Water Act pollutants from its scope (Section 404(a)).

For the Mississippi River, the Section 404 permit program most prominently applies to wetland filling activities along the river and its tributaries. For example, Section 404 permits have been required to build a marine terminal in coastal wetlands and to construct a sewer in wetlands along an Illinois River tributary (*City of Shoreacres v. Waterworth*, 332 F. Supp.2d 992, 1016-17 (S.D. Tex. 2004); *United States v. Hummel*, 2003 WL 1845365, at *4-*5 (N.D. Ill. 2003)).

Wetland loss through dredging, draining, and filling has important implications for water quality in the Mississippi River. However, it is important to remember that most wetland loss in the Mississippi River Basin occurred *before* Congress enacted the 1972 amendments to the Federal Water Pollution Control Act, and Section 404 of the Clean Water Act has almost nothing to say about the restoration of these past wetland losses. Moreover, given the Mississippi River's role in navigation and commerce, the Corps of Engineers has additional authority under *other* federal statutes, most notably the Rivers and Harbors Act of 1899, to engage in navigation-related dredging and construction activities that can continue to deplete Mississippi River wetlands. Therefore, although an environmentally conscious application of Section 404 and the EPA guidelines can help to prevent further wetlands loss and require compensatory mitigation and restoration for any new filling and destruction of wetlands, Section 404 of the Clean Water Act is an insufficient legal vehicle to address the full scope of potential wetland restoration in the Mississippi River Basin.

Section 402 National Pollutant Discharge Elimination System Permit Program

The NPDES permit program governs most point source discharges of pollutants into the nation's waters. Moreover, the Clean Water Act allows states to assume NPDES permitting authority (Section 402(b)), and most states have done so (USEPA, 2006a), subject to EPA oversight (Section 402(b), (c)). Thus, in all but a few states, NPDES permits are for all practical matters *state* permits.

NPDES permit conditions usually depend on technology-based effluent limitations, which the EPA generally establishes (Sections 301, 304), or water quality-based requirements, whichever are more restrictive. Technology-based effluent limitations establish quantitative restrictions on pollutant discharges depending on the kind of pollution-reduction technology available to the discharger's industry for the particular industrial processes the discharger is using (Section 301(b)). Specifically, an "effluent limitation" is essentially a numeric limitation on the amount of a given pollutant that can be discharged at the end of the pipe, measured by quantity, rate, or concentration (Section 502(11)).

Most effluent limitations for discharges from POTWs are based on the reductions of pollutants achievable through secondary treatment, or settling and biological treatment, of the sewage (Section 301(b)(1)). For industrial dischargers, effluent limitations have become progressively more stringent since 1972, and the Clean Water Act now requires that effluent limitations for most pollutants, including toxic pollutants, be based on "the best available technology economically achievable," or BAT (Section 301(b)(2)).

Several other requirements also dictate the terms of NPDES permits. For example, many new point source dischargers must comply with the relevant new source performance standards (NSPS) (Section 306). NSPS are technology-based standards that "reflect the greatest degree of effluent reduction which the Administrator determines to be achievable through application of the best available demonstrated technology" (BADT)—"including, where practicable, a standard permitting no discharge of pollutants" (Section 306). As such, NSPS often are more stringent than BAT-based effluent limitations. Similarly, the EPA administrator may set effluent standards for toxic pollutants that are more stringent than the BAT-based effluent limitations, up to and including a complete prohibition on discharges of certain toxic pollutants from specific industrial processes (Section 307(a)). Finally, all dischargers are also subject to inspection, monitoring, recordkeeping, and reporting requirements (Section 308).

In a typical NPDES permit, technology-based effluent limitations dictate the majority of the discharge requirements for the point source. However, if the discharge "would interfere with the attainment or maintenance of that water quality in a specific portion of the navigable waters which shall assure protection of public health, public water supplies, agricultural and industrial uses, and the protection and propagation of a balanced population of shellfish, fish, and wildlife, and allow recreational activities in and on the water," the NPDES permit must include more stringent water quality-related effluent limitations to ensure that these uses are protected (Section 302(a)). These water quality-based effluent limitations help ensure that the Clean Water Act achieves its overall water quality goals by establishing effluent limitations that meet the water quality standards established by the states for particular waterbodies.

Agricultural Exemptions

Importantly for Mississippi River water quality, the CWA's exempts many agricultural sources of water pollutants from its regulatory scope (Ruhl, 2000). For example, beyond implicitly eliminating nonpoint source pollution from the CWA's regulatory scheme, the act's definition of "point source" *explicitly* excludes "agricultural stormwater discharges and return flows from irrigated agriculture" (Section 502(14)). The Clean Water Act reinforces this exemption by prohibiting the EPA administrator from requiring an NPDES permit "for discharges comprised entirely of return flows from irrigated agriculture, nor shall the Administrator directly or indirectly, require any State to require such a permit" (Section 402(*l*)). Finally, the act explicitly excludes from the dredge-and-fill permit requirement any discharges of dredged or fill material: (1) "from normal farming, silviculture, and ranching activities such as plowing, seeding, cultivating, minor drainage, harvesting for the production of food, fiber, and forest products, or upland soil and water conservation practices"; (2) "for the purpose of construction or maintenance of farm or stock ponds or irrigation ditches, or the maintenance of drainage ditches"; and (3) "for the purpose of construction or maintenance of farm roads . . . , where such roads are constructed and maintained in accordance with best management practices" (Section 404(f)). However, the Clean Water Act's definition of point source *does* explicitly include "concentrated animal feeding operations," or CAFOs, as sources whose discharges of pollutants are subject to the act (Section 503(14)). Thus, discharges from these large-scale "factory farms" *are* regulated under the Clean Water Act, generally through NPDES permits (Ruhl, 2000).

STATE-LEVEL AUTHORITY IN PROTECTING WATER QUALITY

Water Quality Standards

As discussed above, beginning in 1965, the FWPCA took a state-focused, water quality standards approach to improving the nation's water quality. The virtues of this approach are local control and flexibility, which the states fought to retain in the 1972 amendments that created the contemporary Clean Water Act, despite considerable evidence that a pure water quality standards approach was not improving water quality (Houck, 1999). These state demands, coupled with environmentalists' demands for a regulatory backstop and congressional need to reconcile the Senate and the House approaches to regulatory reform, led to Section 303 of the transformed Clean Water Act (Houck, 1999). Section 303 remains a significant element of contemporary water quality regulation, primarily because of the TMDL provisions added to the pre-1972 water quality standards approach (Houck, 1999). Box 3-1 contains further discussion of the TMDL program and process.

BOX 3-1
The TMDL Program and Process

The Total Maximum Daily Load program was created by the 1972 Federal Water Pollution Control Act Amendments (now the Clean Water Act) and today is the foundation for the nation's efforts to meet state water quality standards. TMDL represents the amount of a pollutant that can be discharged daily into a waterbody consistent with applicable water quality standards. The TMDL process refers to the process whereby the TMDL is developed and implemented. Failure to meet water quality standards is a major concern across the United States. Roughly 21,000 river segments, lakes, and estuaries have been identified by states as being in violation of one or more water quality standards.

The TMDL program is mandated by Section 303(d) of the Clean Water Act. The TMDL process was little utilized during the 1970s and 1980s, as states focused on bringing point source polluters into compliance with permit limits issued as part of the National Pollutant Discharge Elimination System. Although the NPDES program has been successful on many counts, it has not achieved the CWA water quality goals of fishable and swimmable waters, largely because discharges from other unregulated nonpoint sources of pollution have not been reduced sufficiently. Today, nonpoint sources of pollution (e.g., those involving discharges of nutrients and sediments from agricultural operations) are jeopardizing water quality and shifting the focus of water quality management among the states from effluent-based to ambient water quality-based regulations and other controls.

Under the 1972 amendments, the states retain their pre-1972 authority to set water quality standards for the waters within their borders. "A water quality standard shall consist of the *designated uses* of the navigable waters involved and *water quality criteria* for such waters based upon such uses. Such standards shall be such as to protect the public health or welfare, enhance the quality of water and serve the purposes of this chapter. Such standards shall be established taking into consideration their use and value for public water supplies, propagation of fish and wildlife, recreational purposes, and agricultural, industrial, and other purposes, and also taking into consideration their use and value for navigation" (Section 303(c)(2); emphasis added).

Thus, state water quality standards (which consist of designated use and water quality criteria) establish ambient water quality goals that a given waterbody is required to achieve. The standards represent social and economic considerations, as well as physical, chemical, and biological considerations (NRC, 2001). Designated uses specify the uses that the state wants the body of water to be able to support. Water quality criteria

Under TMDL regulations promulgated in 1992, EPA requires states to list waters that are not meeting water quality standards, which establish both criteria and designated uses for waterbodies. For each impaired waterbody, the state must identify the amount by which the point and nonpoint source discharges must be reduced for the waterbody to attain its stated water quality standards. A TMDL is developed for these listed waterbodies. The TMDL determines a numerical quantity that will not violate state water quality standards, allocating that load among point and nonpoint sources (the Waste Load Allocation [WLA] and the Load Allocation [LA], respectively). The final step in the process is implementation of the TMDL, with the objective of delisting the waterbody. Creating required TMDLs has become one of the pressing water quality challenges for states in the last two decades.

Water quality monitoring and data play a key role in the TMDL process. They are important in the decision to add waters to the list of impaired waterbodies required by Section 303(d) and in developing TMDLs for listed waters. They are also important in allocating pollutant loads among point and nonpoint sources in a given watershed or river basin. The TMDL process is cyclical, as TMDLs are periodically assessed for their achievement of water quality standards including designated uses. If implementation of the TMDL is not achieving attainment of the designated uses, the use of scientific data and information is part of the process for revising the TMDL.

SOURCE: NRC (2001).

specify the characteristics of the water quality necessary to support those designated uses.

The EPA establishes federal reference water quality criteria for states to use in setting their water quality standards (Section 304). These federal criteria usually do not legally bind the states to any particular water quality criteria, and the Clean Water Act leaves the states free to establish their own water quality criteria, as long as those state criteria are either (1) more stringent than the federal water quality criteria or (2) scientifically defensible as protecting the designated uses, even if the state criteria are less stringent than the federal. However, Congress occasionally creates exceptions to the "guidance" status of federal water quality criteria. For example, Congress specified that the EPA's relatively recent bacteria water quality criteria would apply in any state with coastal recreation waters (fresh or salt) that did not adopt its own such criteria by 2004 (Section 303(i)). Box 3-2 discusses these new bacteria water quality criteria and EPA's accommodation of individual state needs and preferences at the federal level.

Nevertheless, although most of the federal water quality criteria legally function merely as guidance for the states—not requirements—states often use them in setting their own water quality standards. These EPA-set criteria reflect the latest scientific knowledge regarding a variety of water quality impacts from many different pollutants, including impacts on human health, various kinds of aquatic life, and aesthetics and recreation. For example, human health-related criteria would be appropriate water quality criteria for waters designated for drinking water use, while criteria based on aquatic toxicity would be appropriate for waters designated to support fishing and aquatic life. Thus, federal water quality criteria play an important role in the development of state water quality standards. Section 303 also requires each state to adopt an antidegradation policy as part of its water quality standards program. The federal antidegradation policy limits a state's discretion to allow the existing condition of its waters to deteriorate and generally prohibits the loss of existing uses (Section 303(d)(4)).

The EPA reviews state-submitted water quality standards for consistency with the Clean Water Act and must promulgate water quality standards for any state that will not either submit its own water quality standards or adequately respond to the EPA's disapproval (Section 303(c)). However, given the Clean Water Act's emphasis on the states' roles in setting water quality standards, such EPA intervention is rare and most states have promulgated their own water quality standards. As noted above, however, the reference water quality criteria that EPA establishes significantly influence state water quality standards.

BOX 3-2
EPA Oversight of Water Quality Standards: Bacteria Criteria for Coastal Recreation Waters

On November 16, 2004, the EPA promulgated bacteria water quality criteria for states to incorporate into their coastal water quality standards (69 Fed. Reg. 67,218). In the Beaches Environmental Assessment and Coastal Health (BEACH) Act of 2000, Congress had required these criteria for any coastal state—including the Great Lakes states—that had designated areas of its coastal waters for recreation (the state-established designated use), setting a deadline of April 10, 2004, for state submissions of bacteria water quality standards. Although 14 coastal states and territories complied with this deadline, 21 others states did not and thus became subject to the EPA-set criteria (69 Fed. Reg. at 67,234).

Several aspects of EPA's bacteria water quality criteria could be relevant to EPA actions for the Mississippi River. First, even though defaulting states are bound by the EPA criteria, in response to requests from New York and Pennsylvania, the EPA promulgated two sets of criteria for different bacteria indicators for the Great Lakes and then allowed the states the flexibility to choose which set to use (69 Fed. Reg. at 67,223). The EPA provided "this flexibility to all Great Lakes States in the rule because the Great Lakes States have a history of cooperating to protect the Great Lakes resource, and may find a need to agree on a consistent pathogen indicator for the Great Lakes" (67 Fed. Reg. at 67,223). Thus, interstate cooperation became relevant to the implementation of federal criteria in interstate waters.

Second, the EPA adopted "single sample maximum" criteria that required state action (notification of the public and potential beach closure) from a single unusually high sample of bacteria-contaminated waters. However, it left states "the discretion to use single sample maximum values as they deem appropriate in the context of Clean Water Act implementation programs [such as NPDES permitting and TMDLs] other than beach notification and closure" (69 Fed. Reg. at 67,225-67,226). Thus, the EPA envisioned a multi-federal-criteria system in which violation of criteria would trigger different state responsibilities.

Third, the EPA included a five-year compliance schedule for existing point source dischargers that would be affected by the new criteria (69 Fed. Reg. at 67,229). Thus, the EPA recognized that immediate compliance with the federal water quality criteria might not be economically realistic and allowed for phased-in implementation.

Fourth, the EPA allowed individual states to avoid the federal bacteria criteria, which were based on *Escherichia coli* and *Enterococci*, when the state already had some other regulatory mechanism that effectively ensured the attainment of the *purpose* of the federal bacteria criteria—to protect human health. Thus, when the State of Washington demonstrated that its own criterion for fecal coliform adequately ensured limits on other bacteria and protected human health, the EPA released the state from the federal requirement (69 Fed. Reg. at 67,231).

Overall, the bacteria water quality criteria demonstrate that the EPA can impose a federal layer of water quality regulation while retaining state flexibility in implementation and sensitivity to concerns of consistency among states in interstate waters.

Nonpoint Source Pollution Programs

As noted, the Clean Water Act does not define "nonpoint source"; instead, by implication, the term refers to any source of water pollutants that is not a point source. Examples of nonpoint source pollution are runoff from a mall parking lot or from agricultural land into a stream. As a result, nonpoint source pollution is not subject to either of the Clean Water Act's two permitting programs and therefore remains a weakly regulated water quality issue: "The United States has made tremendous advances in the past 25 years to clean up the aquatic environment by controlling pollution from industries and sewage treatment plants. Unfortunately, we did not do enough to control pollution from diffuse, or nonpoint sources. Today, nonpoint source pollution remains the Nation's largest source of water quality problems" (USEPA, 2004a).

The Clean Water Act leaves regulation of nonpoint source pollution to the states. Prior to 1987, states addressed nonpoint source pollution, if at all, only through areawide waste treatment management plans (Section 208). Although designed primarily to encourage each state to plan for the construction of POTWs throughout the state, such plans also required states to "identify, if appropriate, agriculturally and silviculturally related nonpoint sources of pollution" and "set forth procedures and methods (including land use requirements) to the extent feasible for such sources" (Section 208(b)). Nevertheless, area-wide management plans were largely considered a failure with respect to effectively addressing nonpoint source pollution.

In 1987 Congress amended the Clean Water Act to establish the nonpoint source management program (WQA, 1987). Under this program, each state had to identify those navigable waters within its boundaries that could not achieve the applicable water quality standards without nonpoint source pollution controls (Section 319(a)(1)). In addition, the state had to identify the significant nonpoint sources contributing pollutants to these waters, describe a process for identifying best management practices (BMPs) and measures to control those sources, and identify existing state and local controls on such sources.

After the states filed their initial nonpoint source management reports with the EPA, the new program required them to develop a state nonpoint source management program (Section 319(b)). Each state program had to meet six requirements, including a schedule of annual milestones for implementing BMP requirements for nonpoint sources (Section 319(b)(2)). States submitting reports and programs that the EPA approved became eligible for federal grants (Section 319(h), (i)). Moreover, coastal states had to coordinate their Clean Water Act nonpoint source management plans with their Coastal Nonpoint Pollution Control Programs established pursuant to the federal Coastal Zone Management Act (16 U.S.C. Section 1455b). Finally, additional information about pollutant sources and their contributions, including nonpoint source airborne contributions to water impairment, can come from the Toxics Release Inventory (TRI) database established through the Emergency Planning and Community Right-to-Know Act (42 U.S.C. Sections 11001-11050).

State Reports on Water Quality

The Clean Water Act requires states, beginning in 1976, to prepare and submit to the EPA administrator biennial reports on the quality of navigable waters within their borders (Section 305). These Section 305(b) biennial reports must cover five topics: (1) "a description of the water quality of all navigable waters in such State during the preceding year"; (2) an analysis of the extent to which the navigable waters in the state achieve the act's fishable and swimmable goals; (3) an analysis of the extent to which the Clean Water Act's mechanisms have allowed or will allow the state to achieve the fishable and swimmable goal in all its waters, "together with recommendations as to additional action necessary to achieve such objectives and for what waters such additional action is necessary"; (4) estimates of the environmental impact and the economic and social costs of achieving the Clean Water Act's goals, the benefits such achievement would produce, and an estimated date of achievement; and (5) a description of the role of nonpoint source pollution within the state, with recommendations for controlling such pollution (Section 305(b)(1)). The states' water quality reports are accessible through most state environmental agencies' web sites, and summaries are available from the EPA (USEPA, 2007b).

Federal Consistency with State Water Quality Requirements

The Clean Water Act essentially allows states to veto or condition federally authorized projects that may cause discharges into the state's waters, including the waters of the territorial sea (Section 401). A federal agency cannot issue a permit if the state denies certification (Section 401(a)), and states can condition the certification on specific requirements designed to ensure compliance with the act (Section 401(d)). The U.S. Supreme Court has twice upheld broad state authority to condition federal licenses and permits in order to protect water quality and designated uses, including allowing states to force federally permitted or licensed projects to maintain instream flows for fish and reestablish recreational access to the waters (*PUD No. 1 of Jefferson County v. Washington Department of Ecology*, 511 U.S. 700 (1994); *S.D. Warren Co. v. Maine Board of Environmental Protection* U.S., 126 S. Ct. 1843 (2006)).

Nationally, this state certification requirement can apply to a variety of federal permits and licenses, including hydroelectric dam licenses from the Federal Energy Regulatory Commission (FERC), which were at issue in both *PUD No. 1 of Jefferson County* and *S.D. Warren Co.*; NPDES permits from the EPA in states that still lack delegated permitting authority, which was the subject of the Supreme Court's interstate water pollution decision in *Arkansas v. Oklahoma* (503 U.S. 91 (1992)); and Coast Guard licensing and permitting in coastal waters. However, by far the most common triggers for state certifications along the Mississippi River are applications for Clean Water Act Section 404 permits from the Corps of Engineers for activities occurring in wetlands, followed by Corps of Engineers permits under the Rivers and Harbors Act of 1899 for navigation-related projects (ADEQ, 2006; IADNR, 2006; ILDNR, 2006; KDOW, 2006; LDEQ, 2006; MDEQ, 2006a; MDNR, 2006; MPCA, 2006; TDEC, 2006; WDNR, 2006).

The limited case law that exists indicates that the state certification requirement does not apply to federally permitted or licensed *nonpoint* sources of pollutants. However, the Clean Water Act's nonpoint source management program contains its own consistency provision. This provision requires each relevant federal department and agency to "accommodate . . . the concerns of the State regarding the consistency of such applications or projects with the State nonpoint source pollution management program" (Section 319(k)).

TMDLs and the Legal Intersection of Point and Nonpoint Source Pollution

As discussed previously, states set water quality standards to help achieve ambient water quality goals for a given waterbody. Moreover, the NPDES permitting agency must ensure that the effluent limitations in a particular permit are stringent enough to ensure that the receiving water achieves its water quality standards. The Clean Water Act's primary mechanism for connecting water quality standards, NPDES permit requirements, and nonpoint source regulation is the TMDL analysis. This analysis involves determination of the pollutant loads (mass discharges to the water in a specified period) that a particular waterbody can accept and not exceed the applicable water quality criteria.

At the start of the TMDL process, each state has to "identify those waters within its boundaries for which the [technology-based] effluent limitations . . . are not stringent enough to implement any water quality standard applicable to such waters" and then rank those water quality-limited waters in order of priority, "taking into account the severity of the pollution and the uses to be made of such waters" (Section 303(d)). After generating this Section 303(d) list, the state then sets TMDLs for the specific pollutants that are causing the impairments for each listed waterbody "at a level necessary to implement the applicable water quality standards with seasonal variations and a margin of safety which takes into account any lack of knowledge concerning the relationship between effluent limitations and water quality" (Section 303(d)). The TMDL represents the total amount of a particular pollutant that can be added to the waterbody over a particular period of time without violating the applicable water quality standard.

Permitting agencies then must modify the effluent limitations in NPDES permits to implement the established TMDL (Section 303(d)). Moreover, until the waterbody attains its water quality standards, effluent limitations based on the TMDL "may be revised only if (i) the cumulative effect of all such revised effluent limitations based on the total maximum daily load or waste load allocation will assure the attainment of such water quality standard, or (ii) the designated use which is not being attained is removed in accordance with regulations established under this section" (Section 303(d)).

However, if a waterbody is impaired—that is, does not meet its water quality standards—as a result of nonpoint source pollution, TMDLs can also be a means for encouraging states to address both point and nonpoint sources affecting that waterbody. EPA regulations regarding TMDLs expressly recognize that both point source loadings, termed the *waste load allocation*, and nonpoint source loadings, termed the *load allocation*, are components of the total TMDL (40 C.F.R. Section 130.2(i)). Moreover, the U.S. Court of Appeals for the Ninth Circuit has upheld the EPA's authority to set TMDLs for waterbodies polluted only by nonpoint sources of pollutants (*Pronsolino v. Nastri*, 291 F.3d 1123 (9th Cir. 2002)).

INTERSTATE WATER QUALITY PROTECTION

State Authorities and Responsibilities with Respect to Interstate Water Pollution

As noted above, since long before 1972, interstate water pollution problems have been deemed to be matters for *federal* law and *federal* agencies. Nevertheless, the Clean Water Act provides several interstate-related provisions authorizing state action.

Interstate Compacts and Interstate Agencies Related to Water Quality

Congress explicitly gave its consent to the creation of interstate compacts related to water pollution. Specifically, Congress encouraged states to use interstate compacts for "cooperative effort and mutual assistance for the prevention and control of pollution and the enforcement of their respective laws relating thereto" and to establish interstate agencies to coordinate and enforce interstate regulation (Section 103(b)).

However, unlike the states themselves, these interstate agencies no longer are entitled to federal funding. The Clean Water Act generally authorizes "grants to States and interstate agencies to assist them in administering programs for the prevention, reduction, and elimination of pollution" (Section 106). However, to be entitled to such grants, interstate agencies had to apply to the EPA by early 1973. Thus, interstate agencies created since then to address issues such as Mississippi River water quality are ineligible.

Interstate Considerations in State NPDES Permitting

The state delegation provisions impose interstate obligations on states that choose to issue NPDES permits. Specifically, in order to obtain NPDES permit program authority, each state had "to insure that . . . any other State the waters of which may be affected, receive[s] notice of each application for a permit and to provide an opportunity for public hearing before ruling on such an application" (Section 402(b)). In addition, each delegated state (Section 402(b)) must

> insure that any State . . . whose waters may be affected by the issuance of a permit may submit written recommendations to the permitting State (and the Administrator) with respect to any permit application and, if any part of such written recommendations are not accepted by the permitting State, that the permitting State will notify such affected State (and the Administrator) in writing of its failure to so accept such recommendations together with its reasons for so doing.

Finally, the state issuing the permit must send copies of each permit application, permit action, and permit to the EPA administrator, who can object to the permit's issuance. If the permit-issuing state does not adequately address EPA concerns regarding a downstream state's water quality, the EPA can veto the state permit.

Federal Interstate Authorities and Responsibilities

EPA's General Authority and Duty to Coordinate Transboundary Pollution Regulation

The EPA has multiple sources of authority, and multiple duties, regarding efforts to control transboundary, and especially interstate, water pollution. Most generally, the EPA administrator has three overarching mandatory duties regarding interstate water pollution issues. First, the EPA must "encourage cooperative activities by the States for the prevention, reduction, and elimination of pollution" (Section 103). Second, the EPA must "encourage the enactment of improved and, so far as practicable, uniform State laws relating to the prevention, reduction, and elimination of pollution" (Section 103). Finally, the EPA must "encourage compacts between States for the prevention and control of pollution" (Section 103).

Various other Clean Water Act provisions reinforce these general federal coordination obligations regarding transboundary problems. For example, as noted, Congress gave the EPA responsibility for developing nonbinding guidance water quality criteria and information regarding the implementation of those criteria for states to use in their Clean Water Act programs (Section 304), and the EPA's transboundary responsibilities extend into the international arena (Section 310). Some of the most important of the EPA's interstate authorities and duties are described below.

EPA's Role in Addressing Federally Licensed or Permitted Sources of Interstate Pollution

As discussed above, the Clean Water Act's state certification provisions give states authority to condition federal licenses and permits to ensure that federally licensed activities do not violate state water quality standards and other water quality requirements. However, the certification provision also requires that other states potentially affected by the discharge—generally referred to as downstream states, but for border rivers such as the Mississippi, also including cross-stream states—be given the opportunity to ensure that *their* water quality requirements will be met, as well. Most importantly, "[i]f the imposition of conditions cannot ensure such compliance such agency shall not issue such license or permit" (Section 401(a)). Thus, downstream and cross-stream states are also given an opportunity to object to federal permits and licenses, as are the states in which the proposed discharge will originate. States have increasingly been exercising this authority to protect their water quality.

It is, however, the EPA that represents and evaluates the interests of downstream and cross-stream states' interstate water pollution concerns (see *Arkansas v. Oklahoma*, 503 U.S. 91 (1992)). Specifically, the federal licensing or permitting agency must notify the EPA administrator of the discharge, and "[w]henever such a discharge may affect, as determined by the Administrator, the quality of the waters of any other State, the Administrator within thirty days . . . shall so notify the other State" (Section 401(a)(2)). Affected (downstream or cross-stream) states then have 60 days to determine whether "such discharge will affect the quality of waters so as to violate any water quality requirement in such State"; if so, they can notify the EPA and the licensing agency of objections, and the EPA administrator must then hold a public hearing on the objections (Section 401(a)(2)). The EPA administrator presents the licensing agency with recommendations regarding the affected state's objections. Then, that agency, "based on the recommendations of such State, the Administrator, and upon any additional evidence, if any is presented to the agency at the hearing, *shall* condition such license or permit in such manner as may be necessary to insure compliance with applicable water quality requirements" (Section 401(a)(2)).

EPA's Interstate Oversight of State NPDES Permitting

As noted above, states that assume NPDES permitting authority also acquire interstate obligations to potentially affected downstream and cross-stream states. However, the EPA is the final source of authority in addressing these interstate permitting issues. Ultimately, interstate considerations depend on the EPA's authority to veto state-issued permits that do not consider interstate effects and to take over the permitting process for that NPDES permit.

In keeping with congressional intent, the EPA rarely invokes its veto authority. However, federal courts have upheld the agency's authority to take over the NPDES permitting process to address the concerns of downstream states. Thus, in *Champion International Corp. v. EPA* (850 F.2d 182 (4th Cir. 1988)), when Tennessee complained about an NPDES permit that upstream North Carolina was issuing and negotiations failed to resolve the issue, the U.S. Court of Appeals for the Fourth Circuit upheld the EPA's issuance of an NPDES permit that included terms designed to address Tennessee's concerns.

EPA's Authority to Convene Interstate Nonpoint Source Management Conferences

When upstream nonpoint sources impair downstream water quality and interfere with the attainment of downstream water quality standards, the downstream state can petition the EPA to convene a management conference of all of the relevant states, with the goal of reaching an interstate agreement to regulate the upstream nonpoint sources sufficiently to achieve downstream water quality requirements (Section 319(g)). If the states reach such an agreement, moreover, they must incorporate that agreement into their respective nonpoint source management programs. These nonpoint source management conferences thus could result in the elimination of much state discretion in nonpoint source pollution management. Moreover, through EPA's role and especially as a result of any interstate compact that might arise from the conference, which would have to be approved by Congress, interstate nonpoint source management conferences could effectively federalize this state-based area of water quality management.

Perhaps for these reasons, formal nonpoint source management conferences are not used as often as they might be. Several states, including Alabama, Kentucky, and the Long Island Sound states of New York and New Jersey have made use of the nonpoint source management conference *concept* and have solicited grants from the EPA to do so. One of the few formal Section 319(g) petitions to the EPA, however, was Louisiana's petition regarding the Gulf of Mexico, which the EPA transformed into the Mississippi River-Gulf of Mexico Watershed Nutrient Task Force.

As with the rest of the nation, interstate cooperation to address nonpoint source pollution in the Mississippi River tends to occur outside the CWA nonpoint source provisions. For example, the Mississippi River-Gulf of Mexico Watershed Nutrient Task Force is addressing nutrient pollution caused by both point and nonpoint sources along the entire river (USEPA, 2006a). In contrast, when states such as Louisiana work to address nonpoint source pollution of the Mississippi River through their own nonpoint source management programs, they do not appear to use the Clean Water Act's conferencing mechanism (LDEQ, 1999).

EPA-Led Interstate Management Conferences for the National Estuary Program

In 1987, Congress established the Clean Water Act's National Estuary Program (WQA, 1987). Once an estuary is selected for inclusion in the program because of its national significance, the CWA requires the EPA to hold a management conference in order to assess the overall water quality trends within the estuary, to "develop the relationship between the in place loads and point and nonpoint source loadings of pollutants in the estuarine zone and the potential uses of the zone, water quality, and natural resources," and (Section 320(b)) to

> develop a comprehensive conservation and management plan that recommends priority corrective actions and compliance schedules addressing point and nonpoint sources of pollution to restore and maintain the chemical, physical, and biological integrity of the estuary, including restoration and maintenance of water quality, a balanced indigenous population of shellfish, fish and wildlife, and recreational activities in the estuary, and assure that the designated uses of the estuary are protected.

The National Estuary Program thus provides states and the EPA with a mechanism for comprehensively addressing estuarine water quality, including point, nonpoint, and interstate sources of pollutants. There are seven National Estuaries along the Gulf coast, including the Barataria-Terrebonne Estuary in Louisiana (USEPA, 2006b).

Interstate Implications of EPA-Set TMDLs

State-set TMDLs are predominantly intrastate in focus. For example, the Clean Water Act requires each state to identify water quality-limited waters for "those waters within its boundaries" and to establish TMDLs for those waters (Section 303(d)). Moreover, each state must then use the information generated as part of a continuing planning process within the state (Section 303(e)).

Two kinds of interstate water quality authority issues are relevant to TMDLs. First, a downstream state with impaired waters might attempt to use the TMDL process to directly force particular point and nonpoint sources in upstream or cross-stream states to comply with more stringent discharge limitations and BMP requirements, respectively, in order to help achieve the downstream state's water quality standards. Because sources within the upstream and cross-stream states are the regulatory province of those other states, the Clean Water Act's TMDL provisions probably do not authorize downstream states to engage in this kind of direct cross-border regulation. For example, Florida, which is in the process of establishing a TMDL for mercury in the Everglades, recently implied that it lacked authority to reach out-of-state sources of mercury deposited via the air, even though such cross-boundary atmospheric deposition may be a significant nonpoint source of mercury pollution in the Everglades (FDEP, 2003).

Second, TMDLs must deal with cross-border effects. As noted previously, TMDLs technically have an intrastate focus—the upstream state establishes TMDLs to meet its own water quality standards for its waters. Nevertheless, given that the Clean Water Act, as interpreted by EPA, imposes obligations on upstream states to protect downstream water quality through the adoption of their own water quality standards (40 C.F.R. Section 131.10), Section 303(d) effectively requires an upstream state to adopt a TMDL at a level such that it will prevent interference by its point and nonpoint sources with attainment of downstream state water quality standards. Otherwise stated, in achieving its own water quality standards through compliance with the TMDL, the upstream state will eliminate the downstream effects.

Regardless of an upstream state's interstate TMDL obligations, however, the EPA has the authority to establish TMDLs with both downstream and upstream interstate effects. There are also regulatory requirements (at least for point sources), in the form of more stringent discharge limitations, which are based on water quality criteria developed by the EPA explicitly to address interstate water quality problems. For example, the Clean Water Act requires the EPA to set TMDLs when states fail to do so (Section 303(d)), and the federal courts have upheld the EPA's authority to set federal TMDLs even when only nonpoint source pollutants are contributing to the water quality impairment.

The Clean Water Act also specifies that the EPA must "encourage the enactment of improved and, so far as practicable, uniform State laws relating to the prevention, reduction, and elimination of pollution" (Section 103). TMDLs certainly could be one mechanism for providing such encouragement, especially in combination with EPA-recommended water quality criteria for problematic pollutants.

The EPA increasingly has been asserting its own interstate water quality authority. For example, the EPA has developed a watershed program to encourage states to address water quality issues cooperatively and comprehensively on a watershed basis (USEPA, 2006c; Box 7-2 further discusses EPA's watershed approach to water management). The most active component of the EPA's watershed program thus far is the targeted watershed grant program. Since 2003, the EPA has been funding projects designed to improve the overall water quality, fish productivity, and other qualities of targeted watersheds. Indeed, three targeted watershed projects funded in 2004—the Upper Mississippi River project in Iowa, the Sangamon River project in Illinois, and the Fourche Creek project in Arkansas—were designed specifically to address one of the largest interstate pollution problems: Gulf of Mexico hypoxia caused by Mississippi River pollution. There is also the potential of cross-border water quality trading to implement cross-border TMDLs (Chapter 6 contains further discussion of the water quality trading concept).

As a practical matter, the EPA is already establishing TMDLs that must have interstate regulatory effects if they are to achieve water quality standards. For example, in February 2002, EPA Region 4 set a total mercury TMDL for the Ochlockonee River in Georgia (near its southern border) to satisfy a legal agreement (USEPA Region 4, 2002). Atmospheric deposition of mercury accounts for 99 percent of the mercury loading to the Ochlockonee watershed, and the sources of atmospheric mercury are both local and distant. Thus, achievement of the mercury water quality standard in the Ochlockonee River will require increased regulation of out-of-state sources, probably through the EPA's interstate authority under the Clean Air Act.

Addressing nutrient pollution in the mainstem Mississippi River to improve water quality in the Gulf of Mexico almost would certainly require TMDLs with interstate effects, as Gulf of Mexico TMDLs are already demonstrating. For example, as described in Box 3-3, interstate effects were inevitable when EPA Region 6 and Louisiana developed a mercury TMDL for the Louisiana coastal waters of the Gulf of Mexico.

Finally, EPA's reference water quality criteria must "accurately reflect the latest scientific knowledge . . . on the kind and extent of *all* identifiable effects on health and welfare . . . which may be expected from the presence

BOX 3-3
The EPA-Set Louisiana Coast-Gulf of Mexico Mercury TMDL

In June 2005, EPA Region 6 and the State of Louisiana established a fish tissue mercury TMDL for the coastal bays and Gulf of Mexico waters of Louisiana (USEPA Region 6, 2005). This TMDL necessarily implicates the entire Mississippi River, because "the Mississippi River represents a significant source of [mercury] to the Coastal Bays and Gulf Waters of Louisiana because of the large drainage area and massive flow rate. . . . The total mercury load from the Mississippi River is estimated at 2,117,000 grams per year. Classification of [mercury] loading from the Mississippi River as a nonpoint source is necessary since it was beyond the scope of these TMDLs to differentiate point sources from nonpoint sources of mercury for a geographic area covering almost two-thirds of the continental United States." The sources of mercury pollution on the Louisiana Gulf Coast, and the necessary reductions in mercury loadings from those sources, *including the Mississippi River*, are shown in the following table.

Load Allocations for Coastal Basins

Coastal Segment	Segment Name	Point Source Hg Load (g/yr)	NPS Hg Load (g/yr)	Total Hg Load (g/yr)	Hg Load Reduction (g/yr)	NPS Load Allocation (g/yr)
010901	Afchafalaya Bay and Delta	174	55,629	56,803	32,924	22,705
021102	Barataria Basin Coastal Bays	324	94,590	94,914	56,000	38,591
042209	Lake Pontchartrain Basin Coastal Bays	527	52,188	52,715	31,102	21,086
070601	Mississippi River Basin Coastal Bays	0	2,127,578	2,127,578	1,255,271	872,307
110701	Sabine River Basin Coastal Bays	57	20,077	20,134	11,879	8,198
120806	Terrebonne River Basin Coastal Bays	985	115,321	116,306	68,620	46,700

NOTE: NPS = nonpoint source.

of pollutants in any body of water" (Section 304(a)(1); emphasis added). This broad command certainly extends to interstate water quality effects and the cumulative impacts of pollutants along large rivers such as the Mississippi. Coupled with the EPA's authority to approve and disapprove state water quality standards and to encourage cooperative interstate efforts, this broad grant of water quality criteria-setting authority could allow

Given these contributions and necessary reductions, EPA Region 6 assigned the mercury *waste load allocations* (WLA; point sources), *load allocations* (LA; nonpoint sources), and *margin of safety* (MOS) as follows:

TMDL Summary

Coastal Segment	Segment Name	TMDL (g/yr)	WLA (g/yr)	LA (g/yr)	MOS (g/yr)
010901	Afchafalaya Bay and Delta	22,879	174	22,705	0
021102	Barataria Basin Coastal Bays	38,915	324	38,591	0
042209	Lake Pontchartrain Basin Coastal Bays	21,613	527	21,086	0
070601	Mississippi River Basin Coastal Bays	872,307	0	872,307	0
110701	Sabine River Basin Coastal Bays	8,255	57	8,198	0
120806	Terrebonne River Basin Coastal Bays	47,685	985	46,700	0

Simultaneously, however, EPA Region 6 noted

> the load allocation for the Mississippi River basin accounts for the mercury load from upstream sources in the basin (including point and nonpoint sources). Because of the large geographic scope of the basin and the difficulty in identifying specific sources, EPA has not allocated specific waste loads to point sources in the Mississippi River basin upstream of the TMDL area.

Thus, EPA Region 6 assumed that it had authority to impose a load allocation on the entire upstream Mississippi River basin. Moreover, it assumed that it had further authority to assign specific waste load allocations to upstream point sources to achieve the Gulf of Mexico TMDL, even though, because of the complexity, EPA Region 6 actually chose not to do so.

the EPA to guide multistate attention to interstate water quality issues on large rivers. Moreover, the Clean Water Act expressly directs the EPA to consider scientific information regarding "the concentration and dispersal of pollutants" and effects such as "rates of eutrophication and rates of organic and inorganic sedimentation for varying types of receiving waters" (Section 304(a)(1)), again indicating that Congress wanted the EPA to look broadly—including across state borders—when establishing its water quality criteria.

SUMMARY

The Clean Water Act of 1972 represented a significant change in U.S. water quality regulation in that the emphasis shifted from a focus on state-level water quality standards to a federal permitting scheme according to technology-based or more stringent water quality-based limits for individual dischargers. The Clean Water Act authorizes water quality programs, requires state water quality standards, requires permits for discharges of pollutants into navigable waters, and authorizes funding for publicly owned wastewater treatment works.

The Clean Water Act has been effective in addressing point sources of water pollutants. In particular, its NPDES permit program's technology-based effluent limitations ensure that easily identifiable industrial point sources and POTWs employ effective pollution control technology. Moreover, while states now issue most NPDES permits and engage in a significant portion of enforcement, the NPDES permit program has the additional advantages of being subject to federal and citizen enforcement (Sections 309 and 505). The Clean Water Act also addresses ambient water quality goals for the nation that its regulatory mechanisms are supposed to achieve. Specifically, the CWA requires states to develop water quality standards that consist of designated uses and water quality criteria that define acceptable pollutant levels for the waterbody given those designated uses.

Notably, however, the Clean Water Act addresses nonpoint source pollution only in a limited, indirect manner. This is a crucial difference given the significance of nonpoint source water pollution throughout the nation and its special importance to Mississippi River and northern Gulf of Mexico water quality. The Clean Water Act's nonpoint source provisions depend on the states' political will to adopt and their economic capacity to enforce legally binding management measures to control runoff and other forms of nonpoint source pollution. State nonpoint source management programs, reinforced through the state water quality standard goals and specified TMDLs, could do much to address nonpoint source pollution. This would require states to have sufficient scientific and technological information to enact enforceable nonpoint source pollution control requirements, along with sufficient financial strength and political will to enforce those requirements.

The Clean Water Act provides a regulatory role for interstate agencies in appropriate circumstances (Sections 103 and 106). The Mississippi River presents an opportunity for states to share regulatory authority with one or more interstate water quality regulatory organizations (although the

CWA does not provide for the financing for interstate bodies of this type formed after 1972). Even if the EPA and the states use these interstate water quality mechanisms to improve Mississippi River water quality, however, Mississippi River water quality will continue to suffer from several problems that the Clean Water Act cannot address. Some of these problems derive from statutory choices that Congress potentially could change. For example, many agricultural sources of water pollution are exempt from the Clean Water Act's provisions that regulate point sources. Furthermore, some impairments of Mississippi River water and environmental quality stem from legacy problems that the Clean Water Act is not designed to address. For example, the Clean Water Act has nothing to say about endangered species, invasive species, habitat destruction, or other threats to biological diversity except to mandate the attainment and maintenance of water quality sufficient to support native fish, shellfish, and ecological communities. The statute can be used only indirectly to influence decisions made regarding navigation or flood control activities that can affect water quality, such as lock-and-dam construction or the dredging of navigation channels, in that it gives states the opportunity to condition federally issued permits and licenses. It does not mandate restoration of wetlands filled or otherwise altered long before the act took its current form. Moreover, some water quality issues—notably mercury contamination—derive from atmospheric deposition. Since the Clean Water Act does not authorize direct regulation of air pollution, it can only respond to these types of problems, but not really prevent them. **As a result, the Clean Water Act cannot be used as the sole legal vehicle to achieve all water quality objectives along the Mississippi River and into the northern Gulf of Mexico. Nevertheless, the Clean Water Act provides a legal framework that, if comprehensively implemented and rigorously enforced, can effectively address many aspects of intrastate and interstate water pollution, although the emphasis to date has been predominantly on the former.**

Section 303 of the Clean Water Act requires states or the EPA to develop TMDLs for waters that do not meet water quality standards. TMDLs require regulators to look comprehensively at *all* sources of water pollution—point source, nonpoint source, and background. As a result, the TMDL provisions are becoming and are likely to remain key provisions of the Clean Water Act in finally achieving the goal of all of the nation's waters being at least fishable and swimmable. **For TMDLs and water quality standards to be employed effectively to manage water quality in interstate rivers such as the Mississippi, it is essential that the effects of interstate pollutant loadings be considered fully in developing the TMDL.**

The Clean Water Act assigns most interstate water quality coordination authority to the EPA. The EPA has mandatory duties to "encourage cooperative activities by the States for the prevention, reduction, and elimination of pollution" and to "encourage the enactment of improved and . . . uniform State laws relating to the prevention, reduction, and elimination of pollution" (Section 103). Moreover, the EPA has clear statutory authority to (1) take over from states the setting of water quality standards and TMDLs when state efforts do not comply with the Clean Water Act's requirements; (2) convene, at a state's request, interstate nonpoint source management conferences; (3) convene multistate conferences to develop comprehensive water quality management plans to protect National Estuaries; (4) hold hearings to address interstate pollution caused by federally licensed or permitted activities, including water-based activities permitted by the U.S. Army Corps of Engineers; and (5) veto state NPDES permits and take over the permitting process to ensure that interstate pollution from upstream or cross-stream point sources is adequately addressed. As a result, the EPA has the authority to establish TMDLs with interstate effects and, at least for point sources, regulatory requirements designed to achieve those TMDLs, including water quality criteria set at levels designed to address interstate water quality problems. **The Clean Water Act also encourages the EPA to stimulate and support interstate cooperation to address larger-scale water quality problems. The act provides the EPA with multiple authorities that would allow it to assume a stronger leadership role in addressing Mississippi River and northern Gulf of Mexico water quality.**

4

Implementing the Clean Water Act Along the Mississippi River

Achieving the goals of the Clean Water Act along the entire length of the Mississippi River and into the Gulf of Mexico presents scientific and regulatory challenges similar to those presented by many of the nation's other waterbodies. At the same time, the size and interstate nature of the Mississippi River entail many distinctive administrative and implementation issues and problems. As discussed in Chapter 3, great progress has been made in the control of point source pollution—or the "first stage" of Clean Water Act implementation. Today, along the Mississippi River and across its basin, the more pressing pollutant issues involve management of nonpoint source sediments and nutrients.

A fundamental factor that inhibits effective implementation of the Clean Water Act along the Mississippi River, particularly in efforts to address nonpoint source pollution, is the limited amount of adequate water quality data. Such data are essential for understanding the condition of a given waterbody and for assessing whether or not that waterbody is attaining its designated uses. These data are also crucial in creating Total Maximum Daily Load (TMDL) allocations and in evaluating TMDL effectiveness. The importance of Mississippi River water quality monitoring is discussed further in Chapter 5.

This chapter discusses the multistate nature of the Mississippi River basin, and how this creates unique challenges regarding Clean Water Act implementation and effective water quality management. Cooperation and coordination among the 10 Mississippi River mainstem states has been largely absent over the years. The states generally have focused their attention and resources on water quality monitoring and protection of waterbodies that lie wholly within their respective boundaries. As explained in this chapter, this has contributed to a situation in which the Mississippi River is to a large degree an "orphan" from a water quality monitoring and assessment perspective.

This chapter examines administrative issues and challenges regarding implementation of the Clean Water Act along the interstate Mississippi River. It begins with discussion of the progress in controlling point source pollution and concludes with a focus on efforts to address the more complicated nonpoint source challenges. It discusses the respective roles and responsibilities of federal and state agencies in implementing the Clean Water Act (CWA) along the Mississippi River; the fragmented jurisdictional picture that underlies and affects CWA implementation; the state of water quality assessment along the 10-state Mississippi River corridor; and the development of TMDLs and nutrient criteria for the river.

THE NPDES PROGRAM AND POINT SOURCE CONTROL ON THE MISSISSIPPI RIVER

NPDES Program Implementation

Water quality protection and improvement programs of many of the states bordering the Mississippi River started well before the increasing national environmental consciousness that began in the 1950s and 1960s and before passage of the original Clean Water Act. As explained in Chapter 3, after the Clean Water Act's passage in 1972, the National Pollutant Discharge Elimination System (NPDES) became an important mechanism for reducing Mississippi River point source pollution. Table 4-1 lists the agencies that, in large part, currently administer the NPDES and water quality standard programs for each of the Mississippi River mainstem states.

Along the Mississippi River, NPDES permits have been issued to thousands of industrial, municipal, and other point source dischargers, both large and small. Table 4-2 identifies the "major" Mississippi River dischargers that currently have NPDES permits. Although the Environmental Protection Agency (EPA) Permit Compliance System (PCS) database gives only a fragmentary and not completely up-to-date picture of the status of the permit program, Table 4-2 nevertheless provides an indication of the extent of major point source discharges to the Mississippi River.

NPDES permits impose "best-technology" requirements on point sources and, therefore, constitute one of the principal mechanisms within the Clean Water Act to reduce pollutant discharges into "navigable waters," which are defined very broadly. Although the NPDES program resulted in substantial reduction of pollutant inputs to the Mississippi River (especially sewage-related pollutants as documented below), limited data

TABLE 4-1 Stage Agencies with Principal Clean Water Act Responsibilities

	Primary Agency	Web Site	Predecessor Agencies	Other Agencies Sharing CWA Responsibility[a]
Minnesota	Minnesota Pollution Control Agency	*http://www.pca.state.mn.us/*	Office of Environmental Assistance[b]	None
Wisconsin	Wisconsin Department of Natural Resources	*http://www.dnr.state.wi.us/*	None	None
Iowa	Iowa Department of Natural Resources	*http://www.iowadnr.com/*	Iowa Natural Resouces Council; Iowa Department of Environmental Quality; Iowa Department of Water, Air, and Waste; Iowa Energy Policy Council	None
Illinois	Illinois Environmental Protection Agency	*http://www.epa.state.il.us/*	Illinois Department of Public Health	None
Missouri	Missouri Department of Natural resources	*http://www.dnr.mo.gov/*	None	None
Kentucky	Kentucky Department of Environmental Protection	*http://www.dep.ky.gov/*	Kentucky Water Pollution Control board	None
Tennessee	Tennessee Department of Environment and Conservation	*http://www.state.tn.us/environment*	Department of Health and Environment	Tennessee Wildlife Resources Agency (commercial fishing bans); Tennessee Department of Agriculture (Section 319)

continued

TABLE 4-1 Continued

	Primary Agency	Web Site	Predecessor Agencies	Other Agencies Sharing CWA Responsibility[a]
Arkansas	Arkansas Department of Environmental Quality	*http://www.adeq.state.ar.us/*	Arkansas Water Pollution Control Commission; Arkansas Pollution Control Commission; Department of Pollution Control and Ecology	Arkansas Natural Resources Commission (Section 319)
Mississippi	Mississippi Department of Environmental Quality	*http://www.deq.state.ms.us/*	Mississippi Department of Natural Resources	None
Louisiana	Louisiana Department of Environmental Quality	*http://www.deq.louisiana.gov/portal*	Louisiana Department of Wildlife and Fisheries, Water Pollution Control Division; Office of Environmental Affairs	Louisiana Department of Health and Hospitals, Safe Drinking Water Program

[a]This does not include agencies that share water monitoring and/or testing or other natural resource functions.

[b]Primarily responsible for solid waste management.

inhibit comprehensive analysis of the extent of water quality improvement brought about by the NPDES program. A judgment with regard to the effectiveness of the NPDES program in cleaning up the Mississippi River would be facilitated by data indicating the amounts of pollutants that would likely be discharged from industrial and municipal sources had the program not been enacted. However, there are no such data at this point (USEPA Inspector General, 2004).

Sewage Treatment Under the Clean Water Act

The Clean Water Act's construction grant and revolving loan fund programs have financed the construction and improvement of thousands of publicly owned treatment works (POTWs) nationwide, producing measurable water quality improvements across the nation and in the Mississippi River. By 2000, almost 16,000 POTWs existed in the United States, about

TABLE 4-2 NPDES Permits for Dischargers into the Mississippi[a,b]

	Facility Type	Facility Number	Total Permits
Minnesota	Sewerage Systems	32	117
	General industrial, other	85	
Wisconsin	Sewerage Systems	18	23
	General industrial, other	5	
Iowa	Sewerage Systems	24	81
	General industrial, other	57	
Illinois	Sewerage Systems	65	167
	General industrial, other	102	
Missouri	Sewerage Systems	13	86
	General industrial, other	73	
Kentucky	Sewerage Systems	2	11
	General industrial, other	9	
Tennessee	Sewerage Systems	5	9
	General industrial, other	4	
Arkansas	Sewerage Systems	13	29
	General industrial, other	16	
Mississippi	Sewerage Systems	7	26
	General industrial, other	19	
Louisiana	Sewerage Systems	78	254
	General industrial, other	176	

[a]Data in this table come from EPA's Envirofacts PCS database as of May 21, 2006; *http://www.epa.gov/enviro/html/pcs/adhoc.html.*

[b]Data obtained from various state agencies varied from PCS data. The reason for this discrepancy appeared to be the inclusion in the PCS database of major dischargers to tributaries of the Mississippi.

29 percent of which were found in the 10 Mississippi River states. Box 4-1 lists examples of sewage treatment improvements and other advances in Mississippi River water quality realized under the Clean Water Act. The EPA and the states plan renovation of many existing POTWs, and expect construction of an additional 1,688 POTWs in the near future, more than 20 percent of which will be in the 10 mainstem states (USEPA, 2003a).

At the same time, however, state and local government needs for sewage treatment funds remain high. In 2003, the EPA indicated that state needs for secondary wastewater treatment, advanced wastewater treatment,

BOX 4-1
Clean Water Act-Related Progress on the Mississippi River

Increases in Dissolved Oxygen: Minneapolis-St.Paul. In the past, sewage pollution strongly affected dissolved oxygen concentrations in the Mississippi River. For example, the 100-kilometer reach downstream from Minneapolis-St. Paul was severely polluted with sewage for many decades, and this discharge degraded water quality and depleted dissolved oxygen downstream through Lake Pepin in pool 4 (Wiebe, 1927; Fremling, 1964, 1989). The depletion of dissolved oxygen adversely affected fish and pollution-sensitive organisms (e.g., nymphs of burrowing *Hexagenia* mayflies). To reduce impacts of pollutants and protect human health, the Twin Cities Metropolitan Wastewater Treatment Plant (St. Paul) was built in 1938 and, in response to the CWA, was upgraded from primary to secondary treatment in 1978. Currently, it treats about 80 percent of the wastewater generated in the metropolitan area and daily discharges about 0.85 million cubic meters of treated wastewater into the Upper Mississippi River (D. K. Johnson, 2006, Metropolitan Council, Environmental Services, St. Paul, Minnesota, personal communication) at pool 2, river mile 834.5 (Boyer, 1984). Improvements to the plant in recent decades have reduced effluent concentrations of biochemical oxygen demand and other pollutants. As early as the 1980s, water quality in the river downstream of the Twin Cities had improved, and burrowing mayflies began re-colonizing suitable habitats (Fremling, 1989; Johnson and Aasen, 1989; Fremling and Johnson, 1990).

Reduction of Sewage Inputs: St. Louis. The reach downstream from St. Louis, Missouri, has also been affected by sewage discharges. St. Louis began using the river officially for municipal waste disposal in 1850, when cholera epidemics swept the city (Corbett, 1997). Raw sewage discharge from the City of St. Louis and surrounding areas continued until 1970, when the first of two major treatment plants was opened by the Metropolitan Sanitary District (Corbett, 1997). Water quality downstream has since improved in response to wastewater treatment, and the last large primary treatment facility was upgraded to secondary treatment in 1993 (MDNR, 1994).

Sewage Treatment: Memphis. In 1970, Memphis, Tennessee, was the largest U.S. city with no wastewater treatment, although studies suggested there was only a modest impact on water quality because of the high dilution factor at its location on the Mississippi. It was not until the late 1960s that Tennessee's Division of Stream Pollution Control could convince Memphis to hire a consultant to conduct a sewage needs study. That 1969 study recommended the construction of two primary wastewater treatment plants. The South Treatment Plant opened in 1975; the North Treatment Plant came online in 1977. Moving from no municipal wastewater treatment to secondary treatment constituted the largest impact in terms of reduction in point source pollutants discharged at any location along the Mississippi River.

sewage collection infrastructure, and combined sewer overflow correction totaled $161.9 billion (USEPA, 2003a). The EPA has, however, noted that the focus of POTW infrastructure spending is changing (USEPA, 2003a):

> Since the early 1970s, EPA has documented significant improvements in the treatment of municipal wastewater. It is expected that in the future municipalities will need to focus more on capital renewal (rehabilitation and replacement) of existing infrastructure than on infrastructure improvements measured by increased population served and improved levels of treatment. This is a reasonable progression because much of the Nation's infrastructure has reached, or soon will reach, the end of its design life.

In light of an aging sewage treatment infrastructure, this 2003 report indicates that funding for sewage treatment infrastructure remains an important water quality issue under the Clean Water Act for the Mississippi River states and the nation as a whole. Beyond construction and rehabilitation of sewage treatment infrastructure is the issue of adequate sewage treatment in existing POTWs. Sewage discharges to the Mississippi River, for example, remain a source (albeit a small percentage) of the nutrients that contribute to hypoxia in the Gulf of Mexico (USEPA, 2001).

Another Mississippi River sewage pollution problem is the continued existence of combined sewer overflows (CSOs) and some sanitary sewer overflows (SSOs) as well. SSOs are not permitted under the Clean Water Act and, where they exist, must be remedied. Discharges from CSOs, which can be permitted under the Clean Water Act, derive from older sewer systems that channel *both* sewage and stormwater through POTWs. Heavy rains can cause these systems to overflow, carrying untreated waste and other pollutants into river systems. Along the Mississippi River, CSO problems vary considerably from location to location. For example, Minneapolis has been working to separate sewers from storm drains since 1922. Today, only 5 percent of the city's surface area drains into a combined sewer system, resulting in only eight outfalls that discharge waters from CSOs (City of Minneapolis, undated). In contrast, further down the river, St. Louis has 208 CSO outfalls, many of which discharge directly into the Mississippi (Metropolitan St. Louis Sewer District, 2006).

The development of POTWs, the concomitant reduction of sewage pollution from municipalities, and the mitigation of industrial point source inputs represent significant achievements of the CWA and the NPDES program. Compliance with discharge limits under the NPDES program has not, however, eliminated water quality problems for the Mississippi River, as Mississippi River water quality also is affected by inputs from many nonpoint source pollutants. Both point and nonpoint pollutants therefore must be adequately managed in order to realize water quality standards. As described previously, the Clean Water Act has achieved many successes in addressing point source pollution, but nonpoint source pollution remains a significant water quality management challenge. One impediment to effectively managing nonpoint sources of pollution is nonexistent or inconsistent water quality standards for the pollutant of interest.

MISSISSIPPI RIVER WATER QUALITY STANDARDS

Although the EPA has oversight authority, particularly with regard to interstate water quality, states implement most of the Clean Water Act, including the establishment of water quality standards. For interstate waterbodies such as the Mississippi River, however, this multistate implementation of the Clean Water Act on the same river often undermines the act's effectiveness. In particular, each state develops state water quality standards that reflect and respect its priorities and preferences, but may not adequately protect water quality and aquatic resources of cross-stream and downstream states.

Inconsistencies Among State Water Quality Standards

The Clean Water Act vests significant, although not unlimited, discretion in the states to designate uses for streams and lakes within and along their borders. This discretion, however, is subject to the Clean Water Act's goal of attaining water quality that supports aquatic life and recreation (the "fishable and swimmable" objectives). State water quality standards authority is analogous to zoning, because the setting of those standards involves determination of whether a particular segment of a stream should be usable, for example, for human contact recreation or as a cold water fishery. The states' power to define the quality of water necessary to meet the designated uses through water quality criteria is constrained by EPA's ability to supercede state scientific and technical judgments where appropriate. State-adopted designated uses for its waterbodies *and* the criteria defining the quality of water necessary to meet those uses are, collectively, referred to as a state's water quality standards.

In this legal and technical context, it is almost inevitable that inconsistencies will arise among state-adopted water quality standards for streams and rivers flowing between or through two or more states. Nevertheless, mere inconsistency in state water quality standards is not necessarily problematic, even if the states with inconsistent use designations and water quality criteria are located along the same river or, indeed, share the river as a common boundary. For example, State A may designate the part of the river within its borders as a cold water fishery, requiring a high dissolved oxygen content. Downstream of State A and also on the river, State B may have designated its portion of the river as a warm water fishery, which would require a lower level of dissolved oxygen. In this instance, the respective water quality goals of States A and B are consistent in the sense that neither will interfere with the other's attainment and maintenance. However, this happy coincidence may not always occur. For instance, State A may designate its half of a river for human contact recreation; State C, directly across the river, may designate its portion for sewage discharge receiving waters. Alternatively, State A may be immediately downstream from State C. In either case, the waters of State A may be at risk as a result of the probably less stringent controls required to meet the regulatory regime of State C. This type of situation may arise along the Mississippi River, where 10 states either share common borders or find themselves the recipients of pollutants discharged upriver.

Many groups have examined and considered the differences among the Mississippi River states' water quality standards. For example, the Upper Mississippi River Basin Association (UMRBA) is a regional interstate organization formed by the governors of Illinois, Iowa, Minnesota, Missouri, and Wisconsin to coordinate the states' river-related programs and policies and work with federal agencies with river responsibilities. The UMRBA sponsors programs and studies related to ecosystem restoration, hazardous spills, water quality, floodplain management and flood control, commercial navigation, and water supply. The UMRBA issues reports on these upper Mississippi River issues and has a long-standing interest in water quality, water quality standards, and the Clean Water Act.

An UMRBA water quality task force studied the water quality standards among the upper Mississippi River states of Illinois, Iowa, Minnesota, Missouri, and Wisconsin and issued a report on the topic in 2004. In its report, the task force noted (UMRBA, 2004):

> Differences among the [Upper Basin] states in their implementation of the Clean Water Act are not necessarily problematic. Indeed, the Clean Water Act explicitly confers broad latitude upon the states. While federal regulations require a state to "ensure that its water quality standards provide for the attainment and maintenance of the water quality standards of downstream waters," uniformity of standards and listing decisions is not necessarily the objective. Thus, state actions on shared water bodies should be consistent with this requirement, but need not be identical. Whether the differences on the Upper Mississippi River among the five states' water quality standards afford differing levels of protection requires further evaluation.

Table 4-3 presents a selection of water quality criteria adopted by the mainstem Mississippi River states that apply to the Mississippi River. This table shows many differences that could, under certain circumstances, undercut the ability of at least some states to achieve their water quality standards. In addition, many variations among state water quality stan-

TABLE 4-3 Water Quality Criteria Applicable to the Mississippi River [1]

	Turbidity [2]	Temperature [2]	pH [2]	Dissolved Oxygen [2][5]	Fecal Coliform [2][8]
Minnesota	10 NTU	30°C	$6.5 \leq X \leq 8.5$	5 mg/L	200 col/100 mL [15]
Wisconsin	N/A	[4]	$6.0 \leq X \leq 9.0$	5 mg/L	200 col/100 mL
Iowa	25 NTU [B]	Cannot add 3°C	$6.5 \leq X \leq 9.0$	5 mg/L	N/A
Illinois	N/A	[3]	$6.5 \leq X \leq 9.0$	5 mg/L	200 col/100 mL [13]
Missouri	"substantial visible contrast"	[3]	$6.5 \leq X \leq 9.0$	5 mg/L [7]	200 col/100 mL [12][16]
Kentucky	N/A	31.7°C	$6.0 \leq X \leq 9.0$	5 mg/L [6]	1,000 col/100 mL [9] [12]
Tennessee	No turbidity or color in such amounts or of such character that will materially affect fish and aquatic life (FAL); none that will result in any objectionable appearance (REC)	30.5°C and the maximum rate of change shall not exceed 2°C/hr	$6.0 \leq X \leq 9.0$ (FAL) $6.5 \leq X \leq 9.0$ (REC)	Daily average of 5 mg/L with a minimum of 4 mg/L (specific to ecoregion 73a)	N/A[D]
Arkansas	50 NTU, 75 NTU stormflow	32°C	$6.0 \leq X \leq 9.0$	5 mg/L	1,000 col/100 mL [9] [10]
Mississippi	50 NTU[A]	32.2°C [C]	$6.0 \leq X \leq 9.0$	Daily average of 5 mg/L with an instantaneous minimum of 4 mg/L	200 col/100 mL (May-Oct) 2,000 col/100 mL (Nov-Apr) [11]
Louisiana	150 NTU	Cannot add 2.8°C	$6.0 \leq X \leq 9.0$	5 mg/L	2,000 col/100 mL [14]

[1] Unless otherwise indicated, all water quality criteria come from the individual state regulations and apply specifically to the Mississippi River.

[2] The specific water quality criteria listed for a particular state for a particular pollutant may vary depending on the designated use for a specific segment of the Mississippi River.

[3] Dependent on month.

[4] Dependent on month.

[5] 24-hour minima.

[6] DO shall not be below 4.0 mg/L on any instantaneous reading.

PCBs (24-hour average except where otherwise indicated) [2]	Chlordane (24-hour average except where otherwise indicated) [2]	Phosphorous [2]	Nitrogen [2]
0.014 ng/L	0.073 ng/L	N/A [E]	N/A [E]
0.01 ng/L	0.41 ng/L	N/A [E]	N/A [E]
0.014 µg/L	0.004 µg/L	N/A [E]	N/A [E]
0.015 ng/L	0.003 mg/L [18]	N/A [E]	N/A [E]
0.000045 µg/L [17]	0.00048 µg/L [17]	N/A [E]	N/A [E]
0.000064 mg/L	0.00080 mg/L	[19]	[19]
0.00064 µg/L	0.0080 µg/L	Must not stimulate algal growth, must meet regional goals. Use 0.25 mg/L to interpret narrative criteria along with biological criteria unless other scientifically defensible method is produced	Must not stimulate algal growth, must meet regional goals. Use 0.39 mg/L to interpret narrative criteria along with biological criteria unless other scientifically defensible method is produced
0.4 ng/L	5.0 ng/L	N/A[E]	N/A[E]
0.00035 µg/L	0.0021 µg/L	N/A [E]	N/A [E]
0.01 ng/L	0.19 ng/L	[20]	[20]

continued

TABLE 4-3 Continued

[7] Aquatic life only.

[8] Daily maximum except where otherwise indicated.

[9] This is a year-round maximum. During summer months, primary contact waters have a 200 colonies per 100 mL maximum.

[10] The bacteria standards are divided between primary and secondary contact, with primary contact occurring May 1 through September 30, wherein fecal coliform bacteria are tied to a geometric mean of 200 colonies per 100 mL and a monthly maximum of 400 colonies. Secondary contact values are a geometric mean of 1,000 colonies per 100 mL, with a monthly maximum of 2,000 colonies. Secondary contact values apply October through April in those waters designated for primary contact activities.

[11] For May through October the samples examined during a 30-day period shall not exceed 400 per 100 mL more than 10% of the time. For November through April, the samples examined during a 30-day period shall not exceed 4,000 per 100 mL more than 10% of the time.

[12] Fecal coliform as geometric mean based on minimum of 5 samples taken in a 30-day period.

[13] May through October only.

[14] This is a year-round maximum. During summer months primary contact waters have a 400 col/100 mL maximum.

[15] April 1st through October 31st.

[16] This is for whole body contact category A; criteria for secondary contact is 1,800 col/100 mL.

[17] These values are based on a 4-day average.

[18] For public water supply.

[19] Balancing test for local nutrient problems.

[20] Current nitrogen-to-phosphorus ratios should be maintained.

[A] The turbidity outside the limits of a 750-foot mixing zone shall not exceed the background turbidity at the time of discharge by more than 50 Nephelometric Turbidity Units (NTU).

[B] Strictly speaking, Iowa does not have an ambient water quality criterion for turbidity for surface waters. The 25 NTU referred to in the table is from Iowa's "general use" narrative criteria and applies only to increases in turbidity downstream from point source outfalls.

[C] In addition, the discharge of any heated waters into a stream, lake, or reservoir shall not raise temperatures more than 2.8°C.

[D] *Escherichia coli* rather than fecal coliform criteria.

[E] N/A: No numerical criteria.

dards reflect the early days of Clean Water Act administration, when the EPA did not rigorously review state-adopted standards for CWA compliance. Despite the act's requirements (Section 303 (c)) for triennial review by EPA of water quality standards, the agency often has failed to revisit these standards for their adequacy.

Figure 4-1 displays uses of the Mississippi River as designated by all 10 mainstem states. Figures 4-2 and 4-3 show some of the designated uses of

Minnesota

Human Uses
Domestic Consumption
Recreation
Industrial Consumption
Agriculture
Aesthetic Enjoyment
Navigation

Fish & Wildlife Uses
Aquatic Life
Wildlife

Other Uses
Limited Resource Value Waters

Wisconsin

Human Uses
Public Health & Welfare
Recreational Use
Non-Public Water Supply

Fish & Wildlife Uses
Aquatic Life
Warm Water Sportfish Community

Other Uses
NONE

Iowa

Human Uses
Drinking Water (some portions)
Human Health
Fish Consumption
Primary Contact Recreation

Fish & Wildlife Uses
Warm Water Aquatic Life

Other Uses
NONE

Illinois

Human Uses
General Use (including recreational use)

Fish & Wildlife Uses
General Use (including aquatic life protection)

Other Uses
NONE

Missouri

Human Uses
Drinking Water Supply
Whole Body Contact Recreation
Secondary Contact Recreation
Industrial Process
Irrigation
Livestock Watering
Human Health Fish Consumption

Fish & Wildlife Uses
Protection of Aquatic Life
Wildlife Watering
Human Health Fish Consumption
Wildlife Habitat

Other Uses
NONE

Kentucky

Human Uses
Primary Contact Recreation
Secondary Contact Recreation
Fish Consumption

Fish & Wildlife Uses
Warm Water Aquatic Habitat

Other Uses
Outstanding State Resource Water (parts)

continued

FIGURE 4-1 Designated uses of the Mississippi River.

Arkansas
Human Uses
Domestic Water Supply
Primary Contact Recreation
Secondary Contact Recreation
Industrial Water Supply
Agricultural Water Supply

Fish & Wildlife Uses
Fisheries
Delta Fishery

Other Uses
NONE

Tennessee
Human Uses
Domestic Water Supply (limited sections)
Recreation
Industrial Water Supply
Irrigation
Livestock Watering
Navigation

Fish & Wildlife Uses
Fish & Aquatic Life
Wildlife Watering

Other Uses
NONE

Louisiana
Human Uses
Drinking Water Supply (limited sections)
Primary Contact Recreation
Secondary Contact Recreation

Fish & Wildlife Uses
Fish & Wildlife Propagation
Oyster Propagation

Other Uses
NONE

Mississippi
Human Uses
NONE

Fish & Wildlife Uses
Fish & Wildlife

Other Uses
NONE

FIGURE 4-1 Continued

the states along the upper Mississippi as the states face each other across the common boundary. These figures are especially interesting because they illustrate the many differences in designated uses of the Mississippi River between states on opposite sides of the river.

Under Section 303(c)(4)(B) of the Clean Water Act, the EPA is empowered to prepare its own water quality standards for a state not only when the state submits one that the EPA deems inadequate to meet the statutory requirements, but also when the EPA, on its own initiative, determines "that a revised or new standard is necessary to meet the requirements of" the Clean Water Act. Under Section 303(c)(2)(A), standards must be "such as to protect the public health or welfare, enhance the quality of water and serve the purposes of this Act." Finally, under Section 103(a), the EPA is directed to "encourage cooperative activities by the States for the prevention, reduction, and elimination of pollution" and "encourage the enactment of improved and, so far as practicable, uniform State laws" for

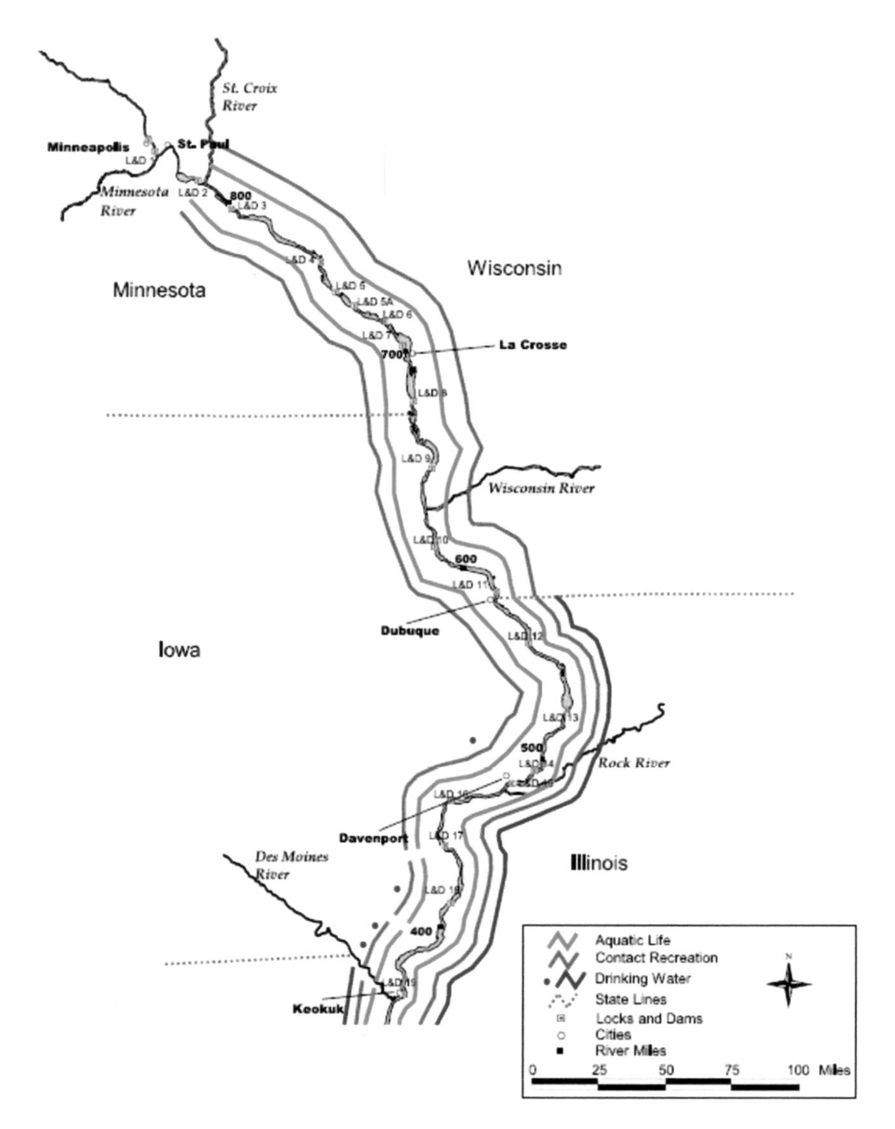

FIGURE 4-2 Designated uses in the upper Mississippi River (part 1 of 2).
SOURCE: Reprinted, with permission from UMRBA (2004). © 2004 by Upper Mississippi River Basin Association.

those purposes. Given that CWA goals, such as the fishable and swimmable mandate, are not in any way limited by the political boundaries that may artificially divide rivers, streams, and lakes, the EPA clearly has the legal authority under these provisions to require states' adoption of uniform

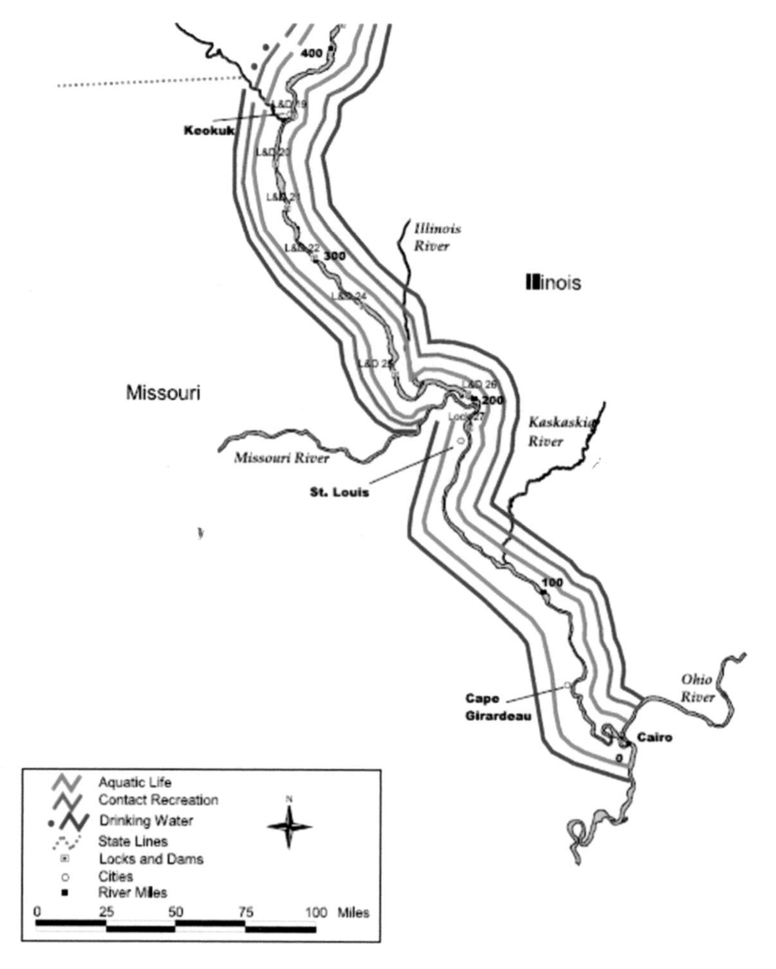

FIGURE 4-3 Designated uses in the upper Mississippi River (part 2 of 2). SOURCE: Reprinted, with permission from UMRBA (2004). © 2004 by Upper Mississippi River Basin Association.

water quality standards, or to impose them if states fail to adopt them, where interstate movement of pollutants may undercut the ability of one or more states to attain the level of water quality they desire. Indeed, the EPA's own regulations require states to take into account such spillover effects: "In designating uses of a water body and the appropriate criteria for those uses, the State shall take into consideration the water quality standards of downstream waters and shall ensure that its water quality standards provide for the attainment and maintenance of the water quality standards of downstream waters" (40 C.F.R. Section 131.10).

The Sierra Club Petition to the EPA

Inconsistencies among Mississippi River state water quality standards have not gone unnoticed. For example, a significant challenge to perceived problems caused by these inconsistencies arose in 2003, when the Ozark Chapter of the Sierra Club submitted a petition to the EPA on February 25 in order to establish adequate and more consistent water quality standards for certain portions of the Mississippi and Missouri Rivers. Specifically, the petition focused on the Mississippi River from Burlington, Iowa, to Memphis, Tennessee. Noting that states along the Mississippi River had listed various segments as not meeting applicable water quality standards and had issued a variety of fish consumption advisories, the petition alleged that designated uses varied up and down and sometimes across the river. The petition asserted, for example, that one upstream state did not designate its portion of the river for drinking water supply or fishing, while certain downstream states did. Also, the underlying criteria for defining use were often not consistent for the same pollutant, even when designated uses were identical. Some states had adopted narrative and others numeric criteria for the same pollutants. Narrative criteria describe the desired water quality goal, such as water free from "objectionable color, odor, or taste"; numeric criteria, by contrast, provide a numerical specification of the required water quality, such as a minimum daily average of 5.0 milligrams per liter of dissolved oxygen (see, for example, Table 4-3). These differences allegedly interfered with the ability of various states to achieve the quality of water identified by their own standards. Moreover, existing standards allegedly were insufficient in some cases to attain the fishable and swimmable objectives of the Clean Water Act. Finally, the petition noted the lack and inconsistency of water quality monitoring among states along the Mississippi River.

On June 25, 2004, EPA refused to take the actions requested by the Sierra Club's petition, although in the process it accepted that, under appropriate circumstances, it had the legal authority under Section 303(c)(4) of the Clean Water Act to grant the type of action requested. However, in deciding in this case whether the promulgation of water quality standards is "necessary to meet the requirements of the CWA," EPA chose to focus on the question of whether the state standards at issue met the "*minimum* requirements of the Act and the federal regulations" in light of the Clean Water Act's intent to preserve the primary role for states in reducing water pollution. Having limited its inquiry in that fashion rather than embracing a more ambitious role for itself in protecting interstate waters, the EPA found that those minimum requirements were indeed met and that the identified inconsistencies did not necessarily undercut protection for adjoining or downstream waters (EPA, 2004b).

As a follow-up to the EPA's action on the Sierra Club's petition, Missouri provided a letter to the EPA committing to adopt no later July 2005 appropriate recreation uses for its segment of the river covered by the petition. However, Missouri's Clean Water Commission ultimately chose not to designate a large portion of the Mississippi River for primary contact recreation, and that decision precipitated a review of the issue by EPA Region 7 (Kansas City). On October 31, 2006, while finding that for 99 of Missouri's internal waterbodies, the state would have to adopt new or revised water quality standards to protect recreational uses, the EPA and the Missouri Coalition for the Environment—which had brought a citizen suit to force the EPA to adopt revised water quality standards for the state—agreed to delay until October 2007 an EPA determination regarding the need for new or revised water quality standards for a 195.5-mile segment of the Mississippi River from the Ohio River to Lock and Dam 27 at Granite City, Illinois (USEPA, 2006d).

WATER QUALITY DATA AND ASSESSMENT FOR THE MISSISSIPPI RIVER

In 2001, a National Research Council (NRC) committee issued a report assessing the TMDL approach to water quality management. A key finding from the report is that "the success of the nation's premier water quality program should not be measured by the number of TMDL plans completed and approved, nor by the number of NPDES permits issued or cost share dollars spent. Success is achieved when the condition of a water body supports its designated use" (NRC, 2001). The NRC (2001) report also notes that a TMDL represents both a planning process to implement standards and a numerical quantity indicating a pollutant load to receiving waterbodies that will not violate state water quality standards with an adequate margin of safety.

Fundamental to creating a TMDL in the first place, and later evaluating its outcomes, are the data that allow one to determine if a waterbody is supporting its designated use, but the information required to make these types of assessments is often not available. Insufficient data prevent the TMDL program—or any water quality improvement mechanism—from adequately evaluating what may be impaired segments of the Mississippi. Adequate water quality monitoring and evaluation are crucial to determining the condition of a waterbody, as well as evaluating outcomes of remedial actions. Good water quality information thus is important to improving water quality in impaired segments of the Mississippi River and bringing them closer

TABLE 4-4 Uses and Assessments of the Mississippi River[a]

State	Designated Use	Percentage of Mississippi River Assessed (river miles)
Minnesota	Aquatic life	68
	Recreation	20
Wisconsin	Aquatic life	100
	Swimming	100
	Fish consumption	100
	General use	100
Iowa	Aquatic life	62
	Drinking water	37
	Primary contact	6
	Fish consumption	61
Illinois	Aquatic life	88
	Drinking water[b]	29
	Primary contact	30
	Secondary contact	0
	Fish consumption	100
Missouri	Aquatic life	100
	Drinking water	100
	Whole body contact	100
	Fish consumption	100
	Boating	100
	Industrial use	100
	Irrigation	100
	Livestock and wildlife watering	100
Kentucky[c]	Aquatic life	100
	Fish consumption	7
	Swimming	16
	Drinking water	0
Tennessee[d]	No Mississippi River Assessments	
Arkansas[e]	No Mississippi River Assessments	
Mississippi[f]	No Mississippi River Assessments	
Louisiana[g]	Primary contact recreation	100
	Secondary contact recreation	100
	Fish and wildlife propagation	100
	Drinking water supply	See note *h*
	Oyster propagation	See note *i*

[a]All data are from UMRBA (2004) unless otherwise indicated.

[b]While Illinois assessed four reaches totaling 29% of its Upper Mississippi River miles for drinking water support, these reaches cover the areas upstream of 9 of the state's 12 public water supplies on the river.

[c]Personal communication with Tom VanArsdall of the Kentucky Department of Environmental Protection on January 12, 2007.

[d]2004 305(b) report.

[e]2002 305(b) report.

[f]2004 305(b) report.

[g]2004 305(b) report.

[h]Assessed from the Old River Control Structure to Monte Sano Bayou (the southernmost tributary to the River located at Baton Rouge) and from Monte Sano Bayou to Head of Passes—includes New Orleans area.

[i]Assessed from the Mississippi River passes to Gulf of Mexico.

to achievement of the CWA fishable and swimmable goal. Moreover, until better information on river quality is obtained and the assessment process is further advanced than it is currently, it will be impossible to implement adequately the Clean Water Act policy to maintain existing good quality water (the so-called nondegradation principle), a policy that all Mississippi River states have adopted into law. Of course, even if data gaps are narrowed or closed, there is no assurance that program administrators will move to the next step of effectively using it for water quality improvement and protection.

The Mississippi River is typical of many of the nation's waterbodies in that current water quality data for all relevant parameters are often unavailable (GAO, 2000a). As recently as 2004, the Upper Mississippi River Basin Association concluded that "[w]ater quality monitoring data on the Upper Mississippi River are currently inadequate for assessing use support and impairments. There are deficiencies in the amount of data, number of monitoring stations, and spatial coverage of existing monitoring" (UMRBA, 2004).

Table 4-4 provides a sense of the degree of water quality assessment to date on the Mississippi River. It indicates that the states along the river differ significantly in the extent of their claimed respective assessments. In considering this information regarding assessed uses, it is essential to note that the method of assessment utilized varies with the state, waterbody, and pollutant of concern. Actual water quality monitoring data may be limited or, in some cases, old or nonexistent, and other means of assessment such as professional judgment may be employed (UMRBA, 2004).

The degree of assessment of the Mississippi River by several states in the lower basin is particularly striking. The availability of water quality information for the lower Mississippi River was raised in one of this committee's meetings (held in Baton Rouge in May 2006). In response to a question regarding identification of the most important water quality issue in the lower Mississippi River, a representative from the Lower Mississippi River Conservation Committee (LMRCC) succinctly answered: "What is the water quality of the Lower Mississippi River?" (Ingram, 2006).

Reasons offered for the dearth of reliable lower Mississippi River water quality data (some of which also apply in the upper basin states) include the following:

- Dangers of sampling in a large, swirling, and fast-moving body of water;
- Limited resources of the states bordering the river;
- A sense that any water quality problems that might be found cannot be solved by one state acting in isolation;
- A state-level focus on intrastate lakes and streams;

MINNESOTA				WISCONSIN		
2002	2004	2006	St. Croix River	2006	2004	2002
(10 segments) 10 PCBs 10 Mercury 4 Turbidity 1 Ammonia Nutrients	PCBs Mercury Turbidity Nutrients	PCBs Mercury Turbidity Nutrients		PCBs Mercury	PCBs Mercury	
			Chippewa River			
(12 segments) 12 PCBs 12 Mercury 1 Fecal coliform	PCBs Mercury Fecal coliform	PCBs Mercury Fecal coliform		PCBs Mercury	PCBs Mercury	
			Lock & Dam 6			
(5 segments) 5 PCBs 5 Mercury 1 Fecal coliform 2 Ammonia	PCBs Mercury	PCBs Mercury	La Crosse	PCBs Mercury	PCBs Mercury	PCBs Mercury
			Root River			
(4 segments) 4 PCBs 4 Mercury 2 Turbidity	PCBs Mercury Turbidity	PCBs Mercury Turbidity		PCBs Mercury	PCBs Mercury	
IOWA						
unlisted	unlisted	unlisted				
unlisted	unlisted					
			Wisconsin River			
unlisted	unlisted	unlisted		PCBs Mercury	PCBs Mercury	
unlisted	unlisted					
			Lock & Dam 11			
unlisted	unlisted	unlisted	Dubuque	PCBs Mercury	PCBs Mercury	
				ILLINOIS		
unlisted	unlisted			PCBs	PCBs	PCBs
						PCBs

continued

FIGURE 4-4 Comparison of impaired waters listings for the upper Mississippi River states, 2002-2006.
NOTES: DO = dissolved oxygen; PCB = polychlorinated biphenyl; TSS = total suspended solids; UMR = upper Mississippi River.
SOURCE: Reprinted, with permission, from UMRBA (2006). © 2006 by the Upper Mississippi River Basin Association.

- Uncertainty of jurisdiction between various states where the river has altered its course over the years;
- Division of authority to manage the Mississippi River among a variety of states and EPA regions; and

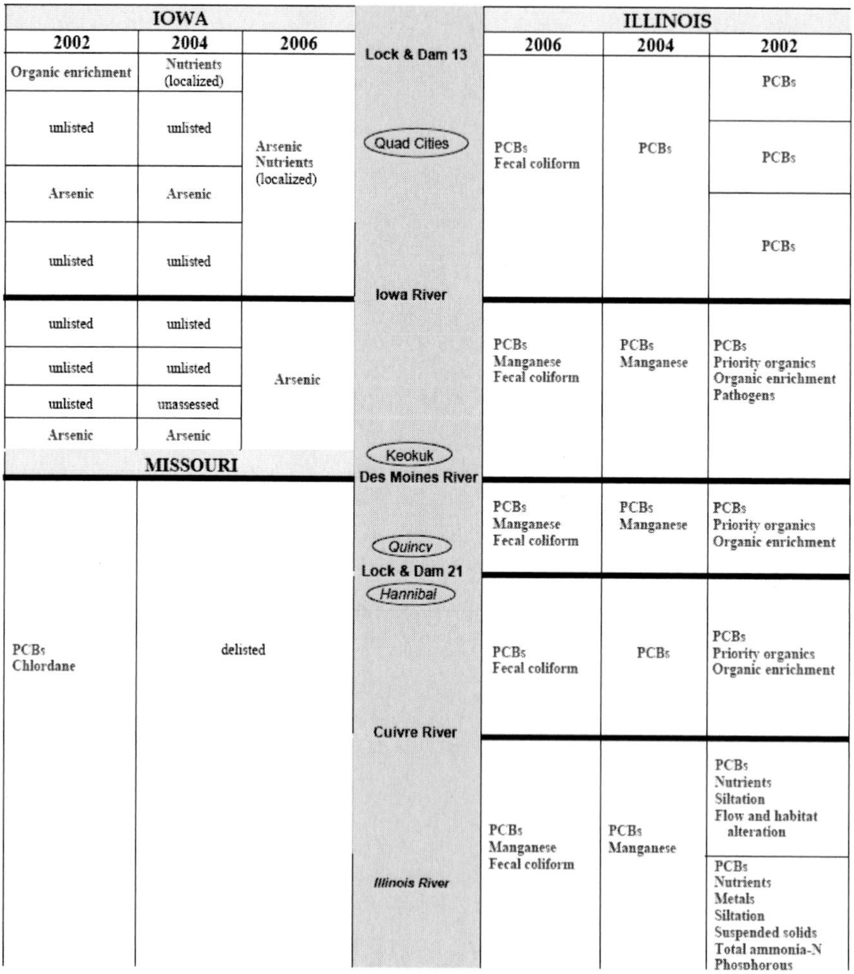

IOWA				ILLINOIS		
2002	2004	2006		2006	2004	2002
Organic enrichment	Nutrients (localized)	Arsenic Nutrients (localized)	Lock & Dam 13	PCBs Fecal coliform	PCBs	PCBs
unlisted	unlisted		Quad Cities			PCBs
Arsenic	Arsenic					
unlisted	unlisted					PCBs
			Iowa River			
unlisted	unlisted	Arsenic		PCBs Manganese Fecal coliform	PCBs Manganese	PCBs Priority organics Organic enrichment Pathogens
unlisted	unlisted					
unlisted	unassessed					
Arsenic	Arsenic		Keokuk			
MISSOURI			Des Moines River			
PCBs Chlordane	delisted		Quincy Lock & Dam 21	PCBs Manganese Fecal coliform	PCBs Manganese	PCBs Priority organics Organic enrichment
			Hannibal	PCBs Fecal coliform	PCBs	PCBs Priority organics Organic enrichment
			Cuivre River			
				PCBs Manganese Fecal coliform	PCBs Manganese	PCBs Nutrients Siltation Flow and habitat alteration
			Illinois River			PCBs Nutrients Metals Siltation Suspended solids Total ammonia-N Phosphorous

continued

FIGURE 4-4 Continued

- Perceptions that the Mississippi River is largely a transportation corridor and that management potential is limited (UMBRA, 2004; Ingram, 2006; Chapter 5 includes additional discussion of the challenges of monitoring a large, interstate river).

The lack of a full and adequate assessment of the Mississippi River for compliance with water quality standards is a crucial issue in considering the

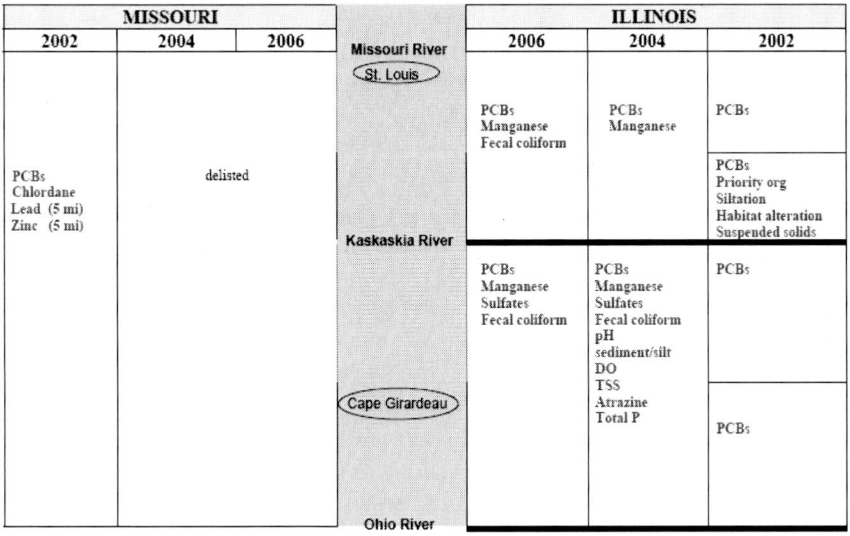

MISSOURI				ILLINOIS		
2002	2004	2006		2006	2004	2002
			Missouri River St. Louis			
PCBs Chlordane Lead (5 mi) Zinc (5 mi)	delisted			PCBs Manganese Fecal coliform	PCBs Manganese	PCBs
						PCBs Priority org Siltation Habitat alteration Suspended solids
			Kaskaskia River			
				PCBs Manganese Sulfates Fecal coliform	PCBs Manganese Sulfates Fecal coliform pH sediment/silt DO TSS Atrazine Total P	PCBs
			Cape Girardeau			PCBs
			Ohio River			

FIGURE 4-4 Continued

Section 303(d) lists of impaired waters (i.e., those not meeting water quality standards that may be candidates for TMDL development) that states must prepare and submit to the EPA biennially. The fact that a particular river segment is not listed as impaired does not necessarily mean that it meets its water quality standards; it may mean simply that the segment has not been assessed or the state does not believe that there is enough information to make a determination of impaired status. More specifically, even if a state claims to have "assessed" a particular river segment for a particular use for the purpose of Section 305(b) (the biennial assessment of the overall quality of a state's waters), that state may not consider the available information sufficient to justify a listing of the river segment as "impaired," an action that could trigger important regulatory obligations, including preparation of a TMDL (UMRBA, 2004). Also, without knowing the length of river segments, the fact that many may be listed says nothing by itself with regard to how much of the river is (or is not) impaired. Indeed, a variety of recent studies have noted the need to use caution in drawing any firm conclusions from state impaired water listings (GAO, 2002; UMRBA, 2004) for a variety of reasons, including the inadequacy of existing water quality standards as measures of ecosystem health and the degree to which states differ in their approaches to impairment assessments and to use and evaluation of available data.

Moreover, in many instances, apparent discrepancies in state listings (for instance, State A lists segment 1 as impaired, while State B across the river does not) result from differences in state water quality standards themselves. Figure 4-4, which is based on the upper states' impaired waters lists, shows differences in these listing among the upper basin states. This figure contains a great deal of information; but it of special interest to note the many differences in these lists between states on opposite sides of the Mississippi River that share the same stretch of the river.

Table 4-5 is based primarily on the 2004 and 2006 Section 303(d) lists and covers both upper and lower Mississippi River states.

An illustration of the apparent anomalies that result from differences in state water quality assessment approaches (e.g., degree of assessment of state waters or data assessment protocols utilized; UMRBA, 2006) is the portion of the Mississippi River that forms the border between Minnesota and Wisconsin. Both states have designated that part of the river for water contact recreation (see Figures 4-2 and 4-3). Moreover, they have similar water quality criteria for fecal coliform bacteria, which is a common parameter describing the suitability of water for such a designated use (see Table 4-3). Yet, for the 2006 Section 303(d) list of impaired waters, Minnesota lists the Mississippi River segment from the Chippewa River to Lock and Dam 6 as impaired for fecal coliform bacteria, while Wisconsin fails to list the same stretch on its side of the river for that parameter (see Figure 4-4).

Several reports on the water quality and ecological integrity of the Mississippi River note sediment (or "siltation") as a priority concern (e.g., USGS, 1999; UMRBA, 2004; Headwaters Group, 2005). However, in contrast to the upper Mississippi River, a lack of sediment replenishment is a crucial problem along the lower Mississippi River and into the Gulf of Mexico. This lack of sediment stems in large part from the dams and reservoirs on the Missouri River, which have trapped vast amounts of sediment that the Missouri previously delivered into the Mississippi River mainstem just above St. Louis. Most states along the Mississippi River have no water quality standards for sediment, although some have turbidity standards and list "siltation" as a cause of water quality impairment. Both Minnesota and Tennessee list the river as impaired with regard to turbidity or siltation (see Figure 4-4). However, Tennessee has no numerical criteria for suspended solids, only narrative criteria, and very few data and no real time-series data to use for such an assessment (P. Davis, Division of Water Pollution Control, Tennessee Department of Environment and Conservation, personal communication, 2006). The absence of sedimentation-related Mississippi River impairments on Section 303(d) lists from several mainstem states can be attributed to a variety of reasons. For example, Iowa has not listed the upper Mississippi River for suspended sediment, sedimentation, or turbidity because the problems associated with those parameters do

TABLE 4-5 Impaired Mississippi River Waters and Pollutants by State

State	Pollutant	Segments Impaired	Mileage of Impairments	Percentage of Mississippi Impaired Within State
Minnesota[a]	Mercury	4	Currently unable to tell, Minnesota doesn't list mileage for impaired segments	
	Fish consumption advisory—mercury	45		
	Fish consumption advisory—PCBs	17		
	Low oxygen	3		
	Turbidity	8		
	Fecal coliform	7		
	Nutrients	1		
	Fish index of biological integrity (IBI)	1		
Wisconsin[b]	Mercury	6	231	100%
	PCBs	6	231	10000%
Iowa[c]	Arsenic	2	162	52%
	Nutrients	1	89	29%
Illinois[d]	PCBs	8	585	81%
	Fecal coliform	7	527	73%
	Manganese	5	348	48%
	Sulfates	1	117	16%
Missouri[e,f]	Lead	1	5	1%
	Zinc	1	5	1%
Kentucky[g]	No impaired segments of the Mississippi River			
Tennessee[h]	PCBs	5	194.4	91%
	Dioxin	5	194.4	91%
	Chlordane	5	194.4	91%
	Nitrate	5	194.4	91%
	Siltation	5	194.4	91%
	Habitat alterations	5	194.4	91%
Arkansas[i]	No impaired segments of the Mississippi River			
Mississippi[j]	Mississippi considers Mississippi River a national task, No 303(d) listings			
Louisiana[k]	Bacteria	1	Currently unable to tell, Louisiana doesn't list mileage for impaired segments	

[a]2006 draft 303(d) report. This includes the intrastate reaches of the Mississippi that are not included in the 2004 UMRBA report.

[b]2006 draft 303(d) report.

[c]2004 303(d) report.

[d]2006 303(d) report.

[e]The earliest expected date of completing the 2006 303(d) report is November 2007 (*http://www.dnr.mo.gov/env/wpp/cwforum/adv-wqm.htm*).

[f]Corrected 2002 303(d) report. However, in the 2006 draft 303(d) report, there are no listed impairments on the Mississippi.

[g]2004 303(d) report.

[h]2004 303(d) report.

[i]2004 303(d) report.

[j]2006 draft 303(d) report.

[k]2006 303(d) report.

not constitute violations of numeric or narrative criteria in Iowa's water quality standards. As a result of all these factors, sediment issues and their complexity in the Mississippi River mainstem are difficult to address from a regulatory standpoint.

THE STATUS OF TMDL DEVELOPMENT ALONG THE MISSISSIPPI RIVER

The mechanism in the Clean Water Act for addressing impairments of water quality by specific pollutants is the development of Total Maximum Daily Loads of the pollutants for the impaired waterbody. In the case of the Mississippi River, impairments have been identified for various segments of the Mississippi River, and a few TMDLs have been developed. Because of resource and other limitations, however, the pace of TMDL development by the 10 Mississippi River mainstem states has been exceedingly slow, more so than with regard to the internal waters of the Mississippi-bordering states and states elsewhere in the nation. Table 4-6 displays the current status for the mainstem of the river. As indicated, while some TMDLs have already been approved by EPA, others have not reached that stage. For comparative purposes, as of March 1, 2007, EPA's TMDL web page listed 64 EPA-approved TMDLs for Minnesota, 136 for Wisconsin, 95 for Iowa, 220 for Illinois, 125 for Missouri, 57 for Kentucky, 641 for Tennessee, 136 for Arkansas, 797 for Mississippi, and 525 for Louisiana (USEPA, 2007c).

Because TMDLs are based on the failure to meet state water quality standards, the applicability of the TMDL process to the Mississippi River depends in part on the stringency of the relevant states' water quality standards—or, more precisely, on the gaps between more or less stringent state water quality standards and actual water quality. If adjacent or cross-stream states do not agree on their water quality standards, their methods of assessing water quality, or both, the TMDL process may apply in one state and not in the other *even though those states are dealing with the same waterbody.*

The current status of the TMDL process on the Mississippi River illustrates the difficulties of imposing this water quality improvement mechanism on a major river bordered by 10 different states. First, states have not even assessed certain segments of the Mississippi River, effectively exempting those segments from the TMDL process for the time being.

Second, states that do list segments of the Mississippi River as impaired waterways pursuant to Section 303(d) do not always agree on how to segment the river or which segments are impaired and for what pollutant(s). For example, as noted earlier, Minnesota lists a stretch of the Mississippi River between the Chippewa River and Lock and Dam 6 as impaired for

TABLE 4-6 TMDL Development Status in the Mississippi River Mainstem States (as of 1/1/07)

State	EPA-Approved TMDLs	EPA-Promulgated TMDLs	State-Adopted TMDLs[a]	Percentage of River in State Covered by TMDLs	Pollutants Addressed
Minnesota	3	0	3[b]	Not available[c]	PCBs, mercury, and nutrients
Wisconsin	0	0	0	0	
Iowa	0	0	0	0	
Illinois	0	0	0	0	
Missouri	1	0	1[d]	100	Chlordane and PCBs
Kentucky	0	0	0	0	
Tennessee	0	0	0	0	
Arkansas	0	0	0	0	
Mississippi	0	0	3[c]	Not available[e]	Legacy pesticides
Louisiana	0	0	0	0	

[a]This number includes both those TMDLs that EPA has approved and those that it has yet to approve but have been completed by the state.

[b]A TMDL for Lake Pepin is not reflected in this number; it is currently being developed.

[c]Minnesota and Mississippi list their Mississippi River segments without mileage. There is no way to determine the mileage covered by TMDLs.

[d]*http://www.dnr.mo.gov/env/wpp/tmdl/wpc-tmdl-pn-mississippi-r.htm*. Covered segments were unlisted but impaired. See *http://iaspub.epa.gov/waters/waters_list.tmdls?state=MO* and Table 4-5.

[e]Personal communication, Richard Ingram, Mississippi Department of Environmental Quality, 1/31/07.

fecal coliform bacteria, while Wisconsin does not list the same segment as impaired for that pollutant on its side of the river. It should be noted that sampling locations for states on opposite sides of the river may be many miles apart, further complicating this issue. That is, if a source of contamination is from an outfall or stream on one side of the river (in Minnesota), coliforms may not reach a sampling location (in Wisconsin) on the other side of the river. Finally, as seen in comparing Tables 4-4 and 4-5 with Table 4-6, states that include segments of the Mississippi River as Section 303(d) impaired waters have not yet established TMDLs for all of those segments. Effective implementation of the TMDL program along the Mississippi River will entail adjustments to the normal state-centered processes of setting water quality standards, delineating river segments, identifying water quality impairments, and resolving significant legal issues.

Development of a TMDL for an interstate river poses several difficulties. For example, again take the case of excessive fecal coliform concentrations on the Mississippi River border of Minnesota and Wisconsin, and assume that both states have listed the same segment as impaired for that pollutant. Each state must prepare a separate TMDL, but the sources of the common problem may be, and probably are, the same. The major point and nonpoint sources of the bacteria may be located equally in both states or situated largely in one or the other, to name but two possibilities. The states should, ideally, work together in identifying the responsible sources and agree on an equitable formula for determining point source waste load allocation and nonpoint load allocation for each state. Moreover, an interstate trading regime might achieve the necessary reductions in the most cost-effective manner, and for these purposes, an appropriate allocation of loads among sources affecting the common resource is absolutely essential. In all events, successful TMDL development and implementation will require close and continuing interstate cooperation and coordination (Chapter 7 discusses U.S. interstate river system compacts and related agreements).

A variation of this Minnesota-Wisconsin example touches on another potential problem of TMDL development in the context of the Mississippi River. Assume, for instance, that on its own, Minnesota opted for a fecal coliform criterion less stringent than Wisconsin's. Also assume that water quality monitoring indicates no violation of the Minnesota standard, but does indicate a violation of the Wisconsin standard. Assume further that Wisconsin has no significant fecal coliform bacteria sources, so the violation of its water contact recreation standard derives entirely from bacteria coming from Minnesota. Must Minnesota design its TMDL to eliminate the violation of Wisconsin's standard? Section 303(d)(1)(A) of the Clean Water Act provides that each state must identify for TMDL development waters "within its boundaries" that do not meet the water quality standards "applicable to such waters." The same apparent intrastate focus is found in Section 303(d)(2) providing for EPA's promulgation of TMDLs for defaulting states.

As noted earlier in this chapter, however, the EPA requires that the water quality standards of a state "provide for the attainment and maintenance of the water quality standards of downstream waters" (40 C.F.R. Section 131.10). Whether a state agency can itself adopt a water quality standard to protect a downstream (or cross-stream) state's water quality is a matter of state law. If the state cannot, the EPA itself would have to adopt the required standard for the state. Therefore, in this example, Minnesota must design its bacteria water quality standard with Wisconsin's fecal coliform standard in mind. If Minnesota's standard is not met, Minnesota must set TMDLs for its portion of the Mississippi River with sufficient stringency to reduce bacteria concentrations that result in downstream violations of Wisconsin's water quality standard. The foregoing scenarios do not depart significantly from cases that have actually arisen. For instance, the ongoing development by the Minnesota Pollution Control Agency of a TMDL for Lake Pepin to deal with eutrophication and turbidity applies to segments of the Mississippi River bordering Wisconsin. That effort has prompted the cooperation of both states, which have differing water quality standards and assessment methods (see MPCA, 2007, for more information regarding Lake Pepin water quality issues and studies).

A further complication is that while TMDLs must be developed even where nonpoint sources are the primary (or sole) cause of water quality standard violations, the creation of TMDLs covering such sources poses significant technical and other challenges not simply for nutrients, but also for other pollutants. This is particularly the case where several states share responsibility for the pollution or are impacted by the pollution, as in the Mississippi River. Even where TMDLs are adopted that cover nonpoint sources, the Clean Water Act does not mandate regulatory controls for them. This fact no doubt inhibits the aggressive development by states of such TMDLs.

Finally, as implemented by the EPA in the past, TMDLs have focused on maintaining longer-term averages. Therefore, they may not be well suited for dealing with "bursts" of storm-driven coliform or nutrient releases, which may be important contributors to Gulf of Mexico hypoxia (Royer et al., 2006). However, a recent case decided by the U.S. Court of Appeals for the District of Columbia Circuit held that, absent additional rulemaking by EPA, the reference to "daily" load in Section 303(d) means daily, not seasonal or annual, load requirements (*Friends of the Earth [FOE], Inc. v. EPA*, 446 F.3d 140 (D.C. Cir. 2006), *cert. denied* 127 S.Ct. 1121 (2007). There is, however, contrary precedent (see *NRDC v. Muszynski*, 268 F.3d 91 (2nd Cir. 2001)). The *FOE* decision, if EPA adheres to it nationally, will ensure that in developing future TMDLs, the EPA and the Mississippi River states will have to deal with short-term nutrient peaks. It bears noting that although the EPA disagrees with the *Friends of the Earth* opinion, it has issued guidance to EPA regions recommending that TMDLs be developed based on daily time increments. It also plans to issue technical documents for deriving daily limits for a variety of pollutants, including nutrients (USEPA, 2006e).

Although development of TMDLs along an interstate river such as the Mississippi River poses various challenges, this Clean Water Act mechanism for addressing water quality impairments can, in many instances, be implemented effectively through cooperation and coordination among the state regulatory entities whose jurisdictions are implicated. The EPA is well positioned to provide this coordination through its authorities under the Clean Water Act and its continuous collaborative efforts with the states in implementing the Clean Water Act. Federal coordination led by the EPA is especially important in addressing large-scale water quality issues such as nutrients and sediments.

NUTRIENT CRITERIA AND TMDLS FOR THE MISSISSIPPI RIVER

None of the 10 Mississippi River mainstem states currently have numeric criteria for nitrogen or phosphorus applicable to the river (listings in Tables 4-4, 4-5, and 4-6 relating to excessive nutrients are based on narrative, not numeric, criteria; UMRBA, 2006). Without such standards, whether they are adopted by individual states or by the EPA, there is little prospect of significantly reducing or eliminating hypoxia in the northern Gulf of Mexico.

Eschewing a "one-size-fits-all" approach in view of the fact that the appropriate concentrations of nutrients (which are necessary for aquatic life) vary with waterbody size, climate, and geology, the EPA has issued guidance for the states to use in developing numeric nutrient criteria (e.g., EPA, 2002). At the same time, the EPA has noted that its recommendations were based on data from smaller waterbodies and that large rivers might require a distinctively different approach (Parker, 2005). As matters stand today, most states are focusing primarily on phosphorus and chlorophyll, and most do not plan to address criteria for large mainstem river systems in the near future (Amy Parker, U.S. EPA, personal communication, 2006).

Even if the Mississippi River mainstem states ultimately develop numeric nutrient criteria for the stretches of the Mississippi River within or on their respective borders, achievement of those criteria would not necessarily resolve the problem of hypoxia in the Gulf of Mexico. To be effective, such criteria would have to be designed specifically with a view to dealing with that large-scale problem. An adequate approach to remediating northern Gulf of Mexico hypoxia would entail establishing numeric nutrient standards for the mouth of the Mississippi and Gulf of Mexico waters that permit no more nutrient flow into the Gulf than could be accommodated by natural processes without significant oxygen depletion. Louisiana, Mississippi, and other Gulf states have the authority to establish such standards to protect their own waters. If they exercise that authority, upstream states will have to create nutrient standards and TMDLs sufficient to achieve the downstream state standards because states must consider the impact of their own water quality standards on waterbodies in adjoining and downstream states. However, the task for upstream states in setting standards with that aggregate downstream effect would require more interstate cooperation and coordination than historically has occurred on the Mississippi River.

In lieu of adequate state action in this situation, the EPA has the legal authority to intervene and create in effect a watershed-wide regime necessary to achieve the same result. At least under certain circumstances, the

EPA's authority under Section 303(c) extends beyond merely harmonizing inconsistent state water quality standards. Under Section 303(c)(4)(B), the EPA can establish a water quality standard "in any case where the Administrator determines that a revised or new standard is necessary to meet the requirements" of the Clean Water Act. Accordingly, the EPA can establish a more demanding standard than any of the states included within a significant national watershed as long as, in the EPA's judgment, that standard is necessary "to restore and maintain the chemical, physical, and biological integrity of the Nation's waters" or to achieve the fishable and swimmable goal of the Clean Water Act. Given Congress's desire generally "to recognize, preserve, and protect the primary responsibilities and rights of States to prevent, reduce, and eliminate pollution" (CWA Section 101(b)), this supervening authority of EPA is most appropriately exercised only in limited circumstances. The Mississippi River, however, would seem clearly to qualify for special treatment, being the nation's only waterbody with congressional recognition as "*a nationally significant ecosystem and a nationally significant commercial navigation system*," as stated in the Upper Mississippi River Management Act of 1986. Moreover, most of the area in the northern Gulf of Mexico that experiences hypoxic conditions is subject to exclusive federal control and protection under the Clean Water Act (see Chapter 3).

Accordingly, the EPA could adopt the necessary numerical nutrient goal(s)(criteria) for the terminus of the Mississippi River and waters of the northern Gulf of Mexico. An amount of aggregate nutrient reduction, from across the entire watershed and necessary to achieve that goal, then could be calculated. Each state in the Mississippi River watershed then could be assigned its equitable share of the reduction. The assigned maximum load for each state then could be translated into numerical water quality criteria applicable to each state's waters.

Each state would then be required to develop a TMDL for "waters within its boundaries" that are identified as failing to meet applicable nutrient criteria, consistent with the language of Section 303(d)(1)(A) of the Clean Water Act. If states failed to adopt the required TMDLs within a reasonable time frame set by the EPA, the EPA could under Section 303(d)(2) promulgate the TMDLs by deeming the failure of states to submit necessary TMDLs a constructive submission of inadequate TMDLs. This "constructive submission" doctrine has so far been developed by the courts as a mechanism to *force* the EPA to act where states have not adopted TMDLs (e.g., *Scott v. City of Hammond*, 741 F.2d 992 (7th Cir. 1984)). Similarly, the EPA could read Section 303(d) in a way that would allow the agency, on its own initiative, to deem a state's failure to act as equivalent to the submission of inadequate TMDLs.

Because TMDL load allocations for nonpoint sources are not legally enforceable under federal law (although states can make them so), and because point sources comprise only a comparatively small percentage (roughly 10 percent) of the nutrient pollutant load transported downstream to the Gulf of Mexico, strong efforts would be required to reduce nonpoint source contributions to the Gulf. In this regard, EPA could, on the petition of Gulf-bordering states or on its own initiative, convene an interstate conference pursuant to Clean Water Act Section 319(g). The conference would be useful in helping reach agreement among Mississippi River watershed states regarding the steps they will take to reduce nonpoint nutrient discharges to meet load allocations established by the nutrient TMDLs.

Improving Mississippi River water quality with respect to nutrients will require coordinated effort among states in TMDL development and other activities on a scale that is commensurate with the scale of the problem. This is a challenge, but there are precedents, most notably from the Chesapeake Bay, where the states in the bay's watershed have been cooperating under EPA leadership for the three-decade-long history of the program.

FEDERAL-STATE COOPERATION IN THE CHESAPEAKE BAY

The case of the Chesapeake Bay offers an example of how the EPA, working collaboratively with the states, can make progress toward nutrient reductions by developing and implementing guidance criteria for new water quality standards for an interstate waterbody. Efforts in water quality improvements in the Chesapeake Bay present an interesting model, with points of comparison and contrast, relevant to the challenges of nutrient loadings into the Mississippi and the Gulf of Mexico.

The Chesapeake Bay is the largest estuary in the United States (Figure 4-5). Its watershed includes parts of six states—Delaware, Maryland, New York, Pennsylvania, Virginia, and West Virginia—and all of the District of Columbia, and drains a basin of 64,000 square miles. From north to south, the bay is approximately 200 miles long; it ranges in width from 3.4 miles in its upstream areas to 35 miles near the mouth of the Potomac River. The bay is relatively shallow, with an average depth of about 21 feet. It supports thousands of species of plants, fish, and animals. More than 16.5 million people live in the Chesapeake Bay watershed area, a figure that is increasing by 1.7 million people every 10 years.

The Chesapeake Bay and its tidal tributaries are listed as impaired waters under Section 303(d) of the Clean Water Act, with nutrients and sediment as the primary sources of impairment. The bay experiences nutrient overenrichment from nitrogen and phosphorus, with pollutant loadings coming from a variety of point and nonpoint sources, including air deposition. Excess nutrients create algae blooms that cloud the water, deprive underwater grasses of sunlight, and consume oxygen that is needed by bay creatures.

FIGURE 4-5 Chesapeake Bay watershed.
SOURCE: Phillips et al. (1999).

Efforts to reduce nutrient loadings to the bay and develop a basinwide, nutrient management program date back to the 1980s. In the late 1970s and early 1980s, Congress funded scientific research on the bay, and the findings pinpointed three areas that required immediate attention: nutrient overenrichment, dwindling underwater bay grasses, and toxic pollution. Once this initial research was completed, the Chesapeake Bay Program was established in 1983 as a regional partnership to direct bay restoration. The program was formed via the Chesapeake Bay Agreement of 1983, which was signed by the governors of Maryland, Virginia, and Pennsylvania; the mayor of the District of Columbia; and the administrator of the U.S. Environmental Protection Agency. Since the signing of the 1983 agreement, the Chesapeake Bay Program partners have adopted two additional agreements that provide overall guidance for bay restoration: the 1987 Chesapeake Bay Agreement and Chesapeake 2000 (C2K). The 1987 agreement established the program's goal of a 40 percent reduction in the amount of nutrients—primarily nitrogen and phosphorus—that enter the bay by the year 2000. The 2000 agreement was signed by the governors of Maryland, Pennsylvania, and Virginia; the mayor of the District of Columbia; the chair of the Chesapeake Bay Commission; and the EPA administrator. The 2000 agreement is being used to guide restoration activities throughout the bay's watershed through 2010. In addition, Delaware, New York, and West Virginia have signed a six-state memorandum of understanding to "work cooperatively to achieve the nutrient and sediment reduction targets that we agree are necessary to achieve the goals of a clean Chesapeake Bay by 2010, thereby allowing the Chesapeake and its tidal tributaries to be removed from the list of impaired waters" (Chesapeake Bay Memorandum of Understanding, 2000. For more information on the Chesapeake Bay Program, see *www.chesapeakebay.net*).

The Chesapeake Bay Program represents a multistate, science-based, cooperative effort, with several different agreements, strategies, and timelines, to reduce nutrient loadings to the bay. Some of the program's prominent aspects follow:

- Multiple states' agreement on shared water quality problems;
- An interstate information management system;
- Basinwide, coordinated monitoring programs and interstate networks;
- A multijurisdictional framework for reporting ecological indicators;
- An agreement on designated uses for shared tidal waters;
- Consistent water quality standards agreed to by upstream states;
- Major tributary basin cap load allocations; and
- A basinwide permitting strategy that addresses 467 facilities.

Figure 4-6 provides further detail of key program components and their relationships.

A key element of the 2000 agreement and the six-state memorandum of understanding is a commitment by Chesapeake Bay watershed jurisdictions to determine the nutrient and sediment load reductions necessary to achieve water quality to protect aquatic living resources. In April 2003, New York, Pennsylvania, Maryland, Virginia, West Virginia, Delaware, the District of Columbia, and the U.S. EPA agreed on the required load reductions that were allocated to each of the watershed's nine major tributary basins and jurisdictions in the form of "cap loads." These cap loads are defined as the maximum amounts of pollutants allowed to flow into a waterbody and still ensure achievement of state water quality standards.

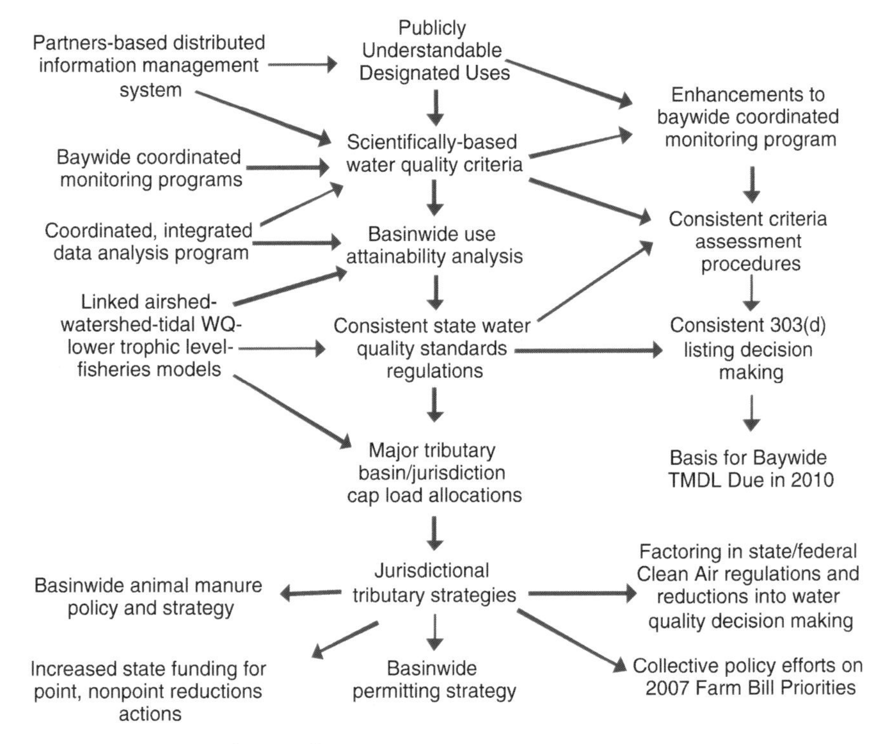

FIGURE 4-6 Key Chesapeake Bay Program components.
SOURCE: Batiuk (2007).

Excess nutrient loadings pose problems for the bay's ecosystems by promoting algal growth, which prevents underwater bay grasses from receiving adequate sunlight and also depletes dissolved oxygen. The Chesapeake Bay Program partners conduct joint water quality modeling through the Chesapeake Bay Program office to project load reductions that would eliminate persistent summer low- to no-dissolved-oxygen conditions in the bay's deep bottom waters. Based on model projections, the partners agreed to cap annual nitrogen loads delivered to the bay's tidal waters at 175 million pounds and to cap annual phosphorus loads at 12.8 million pounds. Sediments suspended in the water column pose problems for bay ecology because they reduce the amount of light available to support healthy and extensive underwater bay grass communities. The Chesapeake Bay Program partners also agreed that sediment loads needed to be reduced in order to achieve water quality conditions that protect aquatic resources. Water quality models were used to determine load reductions necessary to improve water clarity. Annual sediment load was ultimately capped at 4.15 million tons per year, and a goal of new underwater bay grass restoration was set at 185,000 acres (Chesapeake Bay Program, 2003).

Final basinwide nutrient cap loads were allocated to the nine major tributary basins (Figure 4-7, first panel). Basin allocations were further divided and assigned to each of the six watershed states and the District of Columbia based on principles of fairness and equity (Figure 4-7, second panel). These principles were a jurisdiction's impact on bay tidal water quality; progress to date; and the benefit derived from a restored Chesapeake Bay and tidal tributaries. Individual states have the option to further subdivide their major tributary basin cap load allocations into 44 state-defined tributary strategy subbasins (Figure 4-7, third panel). Despite nutrient and sediment pollution reduction efforts over the past two decades, only recently—in 2003—did the EPA and the bay states establish bay-specific water quality criteria for dissolved oxygen, water clarity, and chlorophyll *a*, as well as habitat-oriented tidal water designated uses. The new ambient water quality criteria (USEPA, 2003c, 2003d) were developed in accordance with EPA's *National Strategy for the Development of Regional Nutrient Criteria* (USEPA, 1998a). This national guidance document as it applied

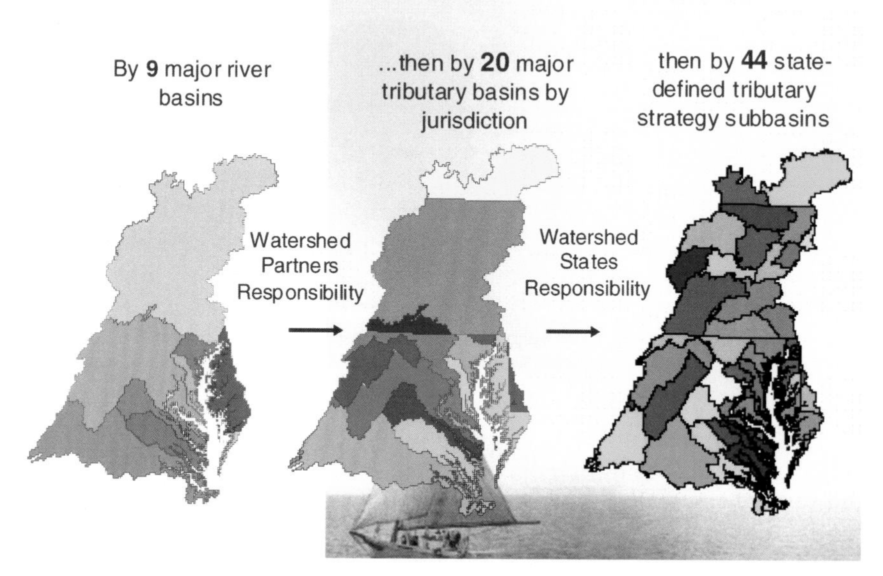

FIGURE 4-7 Chesapeake Bay cap load allocations.
SOURCE: USEPA (2003b).

to the bay was vetted using a multistakeholder approach to implementing Chesapeake 2000. The states, in turn, are incorporating EPA's guidance into their own water quality standards, as both criteria and designated uses, subject to review and approval by EPA, consistent with Clean Water Act requirements.

To date, no formal TMDL has been created for the Chesapeake Bay or its tributaries, although one may be required by court order after 2010 if Chesapeake Bay water quality is not restored by then. Individual states are proceeding with TMDL development for specific waters in the Chesapeake Bay watershed in order to meet agreed-on nutrient and sediment reduction targets. Although there is an existing tributary strategy agreed to by all basin jurisdictions, the water quality criteria now being adopted by each state will be reflected in revised NPDES permits for point source dischargers *for the benefit of the bay and not just local waters*. Specifically, Chesapeake Bay states are moving forward with numerical nitrogen and phosphorus permit limits (annual load limits) for 467 significant municipal and industrial discharges throughout the watershed. EPA is also working on a new permit for the Blue Plains Wastewater Treatment Plant in Washington, D.C., with controls approaching the limits of technology for nitrogen and phosphorus.

In Virginia alone, 125 major dischargers are now required for the first time to reduce nutrients for the benefit of Chesapeake Bay. This development, in turn, prompted the Virginia Legislature to enact a new statute establishing point-to-point source water quality trading under a statewide general permit. Ideally, this will lead to point-to-nonpoint source trading when point sources begin to exceed their allocation caps under the tributary strategy. In addition, Pennsylvania has adopted a nutrient trading policy; it focuses on point-to-nonpoint trading of nutrient loads. Maryland and West Virginia are also developing their own trading policies, and EPA is exploring implementation of an interstate trading regime for that portion of the Potomac River in the Chesapeake Bay basin that encompasses five of the seven jurisdictions. All of these measures face considerable regulatory and technical challenges if they are to be broadened and further developed. Nevertheless, they represent an interest among these states in seeking creative solutions to addressing water quality and nutrient management challenges (Chapter 6 contains further discussion of water quality trading).

With regard to the Chesapeake Bay, there was sufficient interstate consensus for actions that were implemented with a high degree of collaboration. EPA's oversight authority with respect to water quality standards, along with a looming court deadline for a TMDL, provided the impetus for the actions taken. The collaborative efforts among the bay states set a precedent for cooperation in reducing nutrient pollution from sources that do not directly affect local waters.

The Chesapeake Bay Program has experienced some tangible successes to date: during 1990-2000, there was a reported reduction in nutrient loadings to the bay and an increase in the percentage of dissolved oxygen criteria attainment. Wastewater treatment facilities across the watershed also have reported good progress toward reducing nitrogen and phosphorus releases. Nevertheless, the program faces several challenges in its effort to improve water quality and ecological conditions in the bay. The role of agriculture will be especially important, and the program is working with farmers from across the watershed to help meet tillage and conservation goals; substantial progress toward meeting cap loads and water quality goals may well require an unprecedented level of involvement in conservation programs by farming communities. For example, Virginia hopes to see an increase of cropland under conservation tillage from 56 percent in 2002 to 96 percent in 2010 (Batiuk, 2007). In general, there have been some reductions in nutrient loadings from watershed farms, but the current rates of reduction suggest that achievement of restoration goals may still be decades away. The program and the reports of progress on water quality goals have not been without critics. For example, in 2004, the program was accused of overstating its progress toward water quality goals (Washington Post, 2004).

In sum, whether the problem is nutrient pollution of the Chesapeake Bay or the Gulf of Mexico, the value of federal-state and interstate collaboration cannot be overemphasized, especially with regard to adopting and implementing necessary water quality criteria. For Chesapeake Bay, strong interstate and state-federal cooperation, collaboration with municipalities and with the agricultural sector, a thorough scientific process and basis for assessment and for setting goals, and a high degree of transparency have resulted in stronger mutual trust and a comprehensive, coherent nutrient management program across the Chesapeake Bay watershed. The ultimate measure of such programs lies in realizing improvements in water quality and environmental conditions. Given water quality conditions, the administration and implementation of water laws and policies, and land use practices, improvements in water quality will depend strongly on watershed-wide collaborative programs based on effective and consistent water quality monitoring, modeling, and evaluation.

It is worth emphasizing the many years that were required to establish many parts of the Chesapeake Bay program. As mentioned, nutrient loading reduction goals were set in 1987, and the subsequent 20 years saw a lot of give-and-take and numerous meetings and discussions in order to generate the cooperation embodied in the program today. To the extent that the Mississippi River basin states consider the Chesapeake Bay experience in moving forward with basinwide nutrient management programs, this 20-year period should be taken as an indication both of the difficulties involved in such a multistate effort and of the need for immediate, aggressive, and comprehensive action to deal with a pressing environmental problem (in the Gulf of Mexico) of even greater magnitude and complexity.

Fifty years ago the late geographer Gilbert White noted that "no two rivers are the same" (White, 1957). This clearly is the case with the Chesapeake Bay and the Mississippi River and their respective basins. The Mississippi River basin is much larger than the Chesapeake and covers many more states than does the Chesapeake. The Mississippi River also flows through four different EPA regions. At the same time, both basins experience similar water quality problems of excess nutrient and sediment loadings, have a large percentage of land use in agriculture, and administer provisions of the Clean Water Act in a federal-multistate setting. Not all aspects of the Chesapeake Bay Program can necessarily be applied directly to the Mississippi River basin. Nevertheless, the Mississippi River states and the federal government should look to the Chesapeake Bay Program as a useful model in guiding future Mississippi River federal-interstate collaboration on defining and addressing water quality problems, setting science-based water quality standards, and establishing a comprehensive water quality monitoring program.

SUMMARY

The Clean Water Act has provided regulatory mechanisms and financial support that have improved the water quality of the Mississippi River from its pre-1972 condition. In particular, CWA financing of sewage treatment infrastructure construction and the NPDES permit program, with its associated pretreatment requirements for indirect dischargers, have done much to protect Mississippi River water from discharges of raw or partially treated sewage and from industrial wastewater effluent. What the *St. Paul Pioneer Press* reported about local conditions in June 2006 is true for many, though clearly not all, places along the river: "Since the Clean Water Act passed in the early 1970s, more and more people have been reconnecting with a cleaner and more inviting Mississippi River" (St. Paul Pioneer Press, 2006).

Although the Clean Water Act has led to many successes in addressing point source problems, it has not been very effective in addressing large-scale, nonpoint source pollution problems—namely nutrients and sediments—in the Mississippi River. Use of the Clean Water Act to address nonpoint source pollution issues for a large, interstate river such as the Mississippi presents significant challenges. Nonetheless, many key CWA water quality provisions and methods have been under- or poorly utilized in the mainstem Mississippi River. This reflects the river's interstate nature, the expensive and complex task of comprehensively addressing the water quality of the river as an integrated whole, and the inclination of states to divert limited water quality resources to internal waters. Further progress in improving Mississippi River water quality will require improved interstate coordination and cooperation with regard to water quality standards, water quality assessments, TMDLs, and nonpoint source management. Mississippi River states will achieve greater progress in water quality monitoring and other activities by working together, as opposed to each state's working alone. The federal government—namely the EPA—will also have to assume a more aggressive role in Clean Water Act implementation to realize significant Mississippi River water quality improvement.

The Mississippi River serves as a border between states along the length of its corridor running through the middle of the nation. Many states that border the river view Mississippi River water quality as primarily a federal responsibility, and many states allocate only limited funds for water quality monitoring and related activities. Moreover, there is very limited coordination among Mississippi River states in gathering and assessing water quality data and enacting water quality improvement programs. **As a result of limited interstate coordination, the Mississippi River is an "orphan" from a water quality monitoring and assessment perspective.**

Water quality standards differ significantly among Mississippi River states. The Clean Water Act does not necessarily require consistency among state water quality standards. Having uniform standards among all 10 Mississippi River states is neither feasible nor fully necessary for good water quality management. Nevertheless, only the EPA can ensure that a different or less stringent standard of one state does not interfere with the attainment of other states' perhaps more stringent standards.

The Total Maximum Daily Load framework specified in the Clean Water Act has proven useful in managing water quality in some watersheds across the United States, such as the multistate Chesapeake Bay watershed. The TMDL framework, however, presents implementation challenges for large rivers and interstate settings, particularly with respect to nonpoint source pollution. Despite these challenges, the TMDL framework is appropriate for system-wide evaluation of pollutant inputs and for prioritizing control efforts.

The limited degree of interstate coordination and the lack of effective federal oversight, coupled with the failure of many states to actively include the Mississippi River within their state water quality programs, contribute to degradation of water quality in the Mississippi River basin and in the northern Gulf of Mexico. The Clean Water Act *requires* the EPA to oversee and approve state water quality standards and TMDLs; to take over the setting of water quality standards and the TMDL process when state efforts are inadequate; and to *safeguard water quality interests of downstream and cross-stream states.* The Clean Water Act encourages the EPA to stimulate and support interstate cooperation to address larger-scale water quality problems. It also provides the EPA with multiple authorities that would allow it to assume a stronger leadership role in addressing Mississippi River and northern Gulf of Mexico water quality. **The EPA has failed to use its mandatory and discretionary authorities under the Clean Water Act to provide adequate interstate coordination and oversight of state water quality activities along the Mississippi River that could help promote and ensure progress toward the act's fishable and swimmable and related goals.**

The EPA should act aggressively to ensure improved cooperation regarding water quality standards, nonpoint source management and control, and related programs under the Clean Water Act. The EPA is authorized to step in and address water quality problems that may exist because of limited state action in setting and enforcing water quality standards and related Clean Water Act provisions. Indeed, the EPA has the statutory duty to do so. A more aggressive role for EPA in this regard is crucial to maintaining and improving water quality in the Mississippi River and the northern Gulf of Mexico.

There are currently neither federal nor state water quality standards for nutrients for most of the Mississippi River, although standards for nutrients are under development in several states. Both numerical federal water quality criteria and state water quality standards for nutrients are essential precursors to reducing nutrient inputs to the river and achieving water quality objectives along the Mississippi River and for the Gulf of Mexico. A TMDL could be set for the Mississippi River and the northern Gulf of Mexico. This would entail the adoption by EPA of a numerical nutrient goal (criteria) for the terminus of the Mississippi River and the northern Gulf of Mexico. An amount of aggregate nutrient reduction, across the entire watershed, necessary to achieve that goal then could be calculated. Each state in the Mississippi River watershed then could be assigned its equitable share of reduction. The assigned maximum load for each state then could be translated into numerical water quality criteria applicable to each state's waters.

The EPA should develop water quality criteria for nutrients in the Mississippi River and the northern Gulf of Mexico. Further, the EPA should ensure that states establish water quality standards (designated uses and water quality criteria) and TMDLs such that they protect water quality in the Mississippi River and the northern Gulf of Mexico from excessive nutrient pollution. In addition, through a process similar to that applied to the Chesapeake Bay, the EPA should develop a federal TMDL, or its functional equivalent, for the Mississippi River and the northern Gulf of Mexico.

5

Evaluating Mississippi River Water Quality

Accurate evaluation of Mississippi River quality, and how that water quality changes over time, is important for several reasons. This information is essential in measuring the effectiveness of water quality remediation strategies such as Total Maximum Daily Loads (TMDLs). It also is central to determining if water quality standards are being met. More generally, knowledge of water quality in a river or a watershed often is of great interest to citizens, elected officials, and decision makers. Comprehensive and accurate portrayal of water quality conditions requires both the collection of data (monitoring) and an understanding of the system that is supported by scientific investigations (research). Ideally there will be clear and mutually supportive links between monitoring and research. Effective data gathering efforts also require a sustained commitment over time if water quality trends are to be detected and evaluated.

Monitoring and evaluating Mississippi River water quality poses unique challenges because (1) monitoring efforts face logistical difficulties and hazards in some parts of the river system; (2) processes and natural fluctuations in the Mississippi River operate on scales of decades and over hundreds of miles; (3) the river spans, or forms, boundaries of political units or jurisdictions that have differing priorities and resources; and (4) water quality standards and environmental conditions vary across the entire system. For example, because of natural, longitudinal changes in water quality from upstream to downstream, levels of suspended sediment and turbidity that would be considered "pristine" (i.e., pre-settlement) in the lower reaches of the Mississippi River would be considered objectionable and indicative of severe degradation if encountered in the river's headwaters. Likewise, because of natural patterns and differences along the river's length, water quality conditions (e.g., turbidity, temperature, dissolved oxygen) that exist in the headwaters can never be realized in the far downstream reaches. Beyond typical longitudinal patterns, there are also large differences among the subbasins within the Mississippi drainage basin. Any comprehensive evaluation of Mississippi River water quality must consider these differences along the river's length and across the river's watershed.

This chapter examines issues associated with evaluating Mississippi River water quality. It describes some key features of the river and how its hydrologic and watershed characteristics affect water quality monitoring. The chapter reviews past and existing monitoring programs on the Mississippi River mainstem. It discusses the value of river system monitoring in tracking changes in water quality and the importance of monitoring in achieving Clean Water Act goals. It also discusses challenges of using data and information from monitoring programs to help meet Clean Water Act objectives. Finally, this chapter offers recommendations for enhanced state and federal efforts to improve monitoring efficiency, reduce data gaps, and strengthen implementation of the Clean Water Act.

MISSISSIPPI RIVER BASIN STRUCTURE, HYDROLOGY, AND MONITORING

The mainstem Mississippi River exhibits markedly different hydrology, sediment loads, and other features between its upstream and downstream portions. These upstream-downstream differences are driven in large part by inputs from the Mississippi's two main tributaries, the Missouri and Ohio Rivers, which enter the Mississippi at St. Louis, Missouri, and Cairo, Illinois, respectively. The Missouri River is the longest tributary of the Mississippi, and its flow is about two-thirds of the upper Mississippi River above St. Louis. It carries a suspended sediment load several times that of the upper Mississippi River (Meade, 1995). The dams constructed on the Missouri River have reduced the Missouri's total sediment contribution to the Mississippi by more than half since 1953 (Meade and Parker, 1985; Meade et al., 1990). As the Mississippi flows southward, the waters it receives from the Illinois and Missouri Rivers more than double its discharge (Meade, 1995). Downstream, the Ohio River is the Mississippi's largest tributary with respect to discharge, carrying almost twice the discharge of the upper Mississippi River above St. Louis (Table 2-1). Just as the river's discharge doubles when it receives the waters of the Missouri, its discharge more than doubles again as it receives the waters of the Ohio River (Meade, 1995).

Downstream of the Mississippi River's confluence with the Ohio River, the river takes on a very different character than in its upstream reaches. In the Mississippi's lower reaches, the river becomes much deeper and wider and in many areas contains swiftly moving, swirling, and turbulent water. Author John Barry provides a colorful depiction of lower Mississippi River hydraulics in his 1997 book, *Rising Tide*:

> The complexity of the Mississippi exceeds that of nearly all other rivers. Not only is it acted upon; it acts. It generates its own internal forces through its size, its sediment load, its depth, variations in its bottom, its ability to cave in the riverbank and slide sideways for miles, and even tidal influences, which affect it as far north as Baton Rouge. Engineering theories and techniques that apply to other rivers, even such major rivers as the Po, the Rhine, the Missouri, and even the upper Mississippi, simply do not work on the lower Mississippi, which normally runs far deeper and carries far more water.

Monitoring efforts in the river's lower stretches are difficult and hazardous even under relatively calm conditions. These physical differences between the upper and lower Mississippi River influence the ability of the states along the river to monitor water quality and help explain some of the differences in water quality monitoring efforts among the 10 Mississippi River states.

Downstream of Cairo, the influence of direct lateral inputs (i.e., from the adjoining bank or inflowing tributaries) to the Mississippi mainstem becomes relatively less important. In the lower river, water quality thus primarily is a function of upstream inputs, with less influence from the immediately adjacent land. The states of the lower river thus understandably consider the river's condition, and possible water quality remedies, to be largely beyond their control and responsibility. For example, the Mississippi River and its basin upstream of Memphis, Tennessee, represent 80 percent of the total drainage area, 76 percent of the total flow volume, and more than 90 percent of the total riverbank miles for the entire system (Leopold et al., 1964).

A consequence of the structure of the Mississippi River drainage system is that the water quality in the mainstem of the lower river, because of the large and relatively slowly changing mass of water involved, remains relatively constant between those points at which major tributaries join the flow. Thus, closely spaced sampling along the longitudinal axis of the channel generally is not needed to get an accurate measure of average river and water quality conditions over relatively large areas. However, because inflowing tributaries may take many miles to mix completely with the main body of the river, lateral and vertical patterns in water quality can be substantial and persistent. As the following sections explain, the influences of the spatially variable Mississippi River drainage structure on water quality have contributed to differences in U.S. federal and state monitoring of the river and in how states along the river have approached Mississippi River water quality monitoring.

FEDERAL AND REGIONAL MISSISSIPPI RIVER EVALUATIONS

As on many of the nation's large rivers, various types of monitoring have long been conducted on the Mississippi River. River flows have been measured, water quality has been sampled, and ecosystem changes have been tracked. The sum of these monitoring efforts presents a complex and fragmented picture because they have been conducted by different federal and state agencies and scientists, at differing spatial scales and time intervals, with differing objectives, and with varied and changing budgets. Since monitoring efforts are conducted at differing scales and for differing objectives, there is no "one-size-fits-all" or standard river monitoring program.

Monitoring system designs and programs must consider and balance a need for stability and continuity, on the one hand, with changes in scientific paradigms, monitoring technologies and instrumentation, budgets, and political and management objectives on the other. They must cope also with the reality that it is not practical or feasible to monitor continuously every site of interest in the system at hand (e.g., a large river) and that such systems will always contain complexities and unknowns. Scientists must gather and analyze enough information to improve scientific understanding, while recognizing that there are limits to the amount of data that can be gathered and there always will be some uncertainties regarding the state and dynamics of large water systems or ecosystems. To help cope with these realities, models of the system(s) being monitored are often developed so data gathered from individual sites can be used to construct a quantitative or conceptual framework of system-wide dynamics and behavior.

Federal Monitoring Programs

Federal agencies have sponsored and conducted the large-scale monitoring efforts for the Mississippi River. One of today's prominent river monitoring efforts is the Long Term Resource Monitoring Program (LTRMP). Established in 1986 as part of the U.S. Army Corps of Engineers' Environmental Management Program (EMP) for the upper Mississippi River, this initiative seeks to supply essential scientific information to the EMP for the purposes of maintaining the upper Mississippi River as a viable large river ecosystem with multiple uses (USGS, 1999). Since the LTRMP's inception, the Environmental Management Technical Center (EMTC) has implemented the program. The EMTC today is part of the Upper Midwest Environmental Sciences Center, which is a U.S. Geological Survey (USGS) science center. The USGS, the U.S. Fish and Wildlife Service, and the five upper Mississippi River basin states are cooperative partners in the EMP, with the Corps of Engineers responsible for programmatic and financial oversight. The LTRMP samples biota and water quality in five mainstem

FIGURE 5-1 Long Term Resource Monitoring Program study areas (1993-2006). SOURCE: USGS (1999).

reaches upstream of the Ohio River confluence to represent conditions and habitat on the upper Mississippi River system (Figure 5-1). In each LTRMP study reach, several hundred locations have been sampled for biota and water quality since 1993 (Soballe and Fischer, 2004). The LTRMP-EMP issued a comprehensive report in 1999 on upper Mississippi River ecological status and trends. The report was described as "a milestone in the history of the LTRMP. For the first time, data collected since the start of the LTRMP are summarized in one report alongside historical observations and other scientific findings" (USGS, 1999).

In addition to its efforts within the LTRMP, the USGS has been a leader in other Mississippi River monitoring efforts, both in river flows and in water quality sampling. USGS efforts in measuring discharge on the Mississippi River mainstem have remained relatively constant over the years, but there has been a decrease in the extent of its water quality monitoring efforts. For example, at its peak in the 1970s, the National Stream Quality Accounting Network (NASQAN), operated by the USGS, provided extensive coverage of the nation's rivers, including the Mississippi. However, that network has been steadily diminished in the number of sites, the number of samples, and the number of parameters collected, and no other national monitoring programs or monitoring by states and other entities has replaced it. NASQAN data have been useful for several different applications and computations. For example, USGS NASQAN data can be used to compute long-term trends in the monthly nutrient flux at St. Francisville, Louisiana (see Goolsby et al., 1999). Figure 5-2 shows changes in the number of active

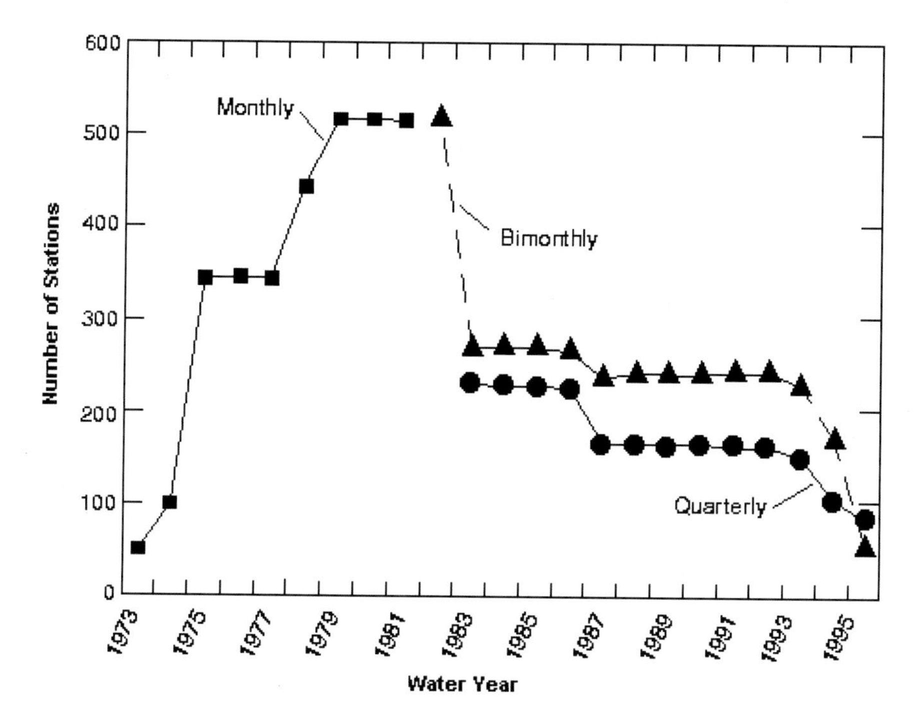

FIGURE 5-2 History of active NASQAN sites at the national level. Reductions in the network that were implemented in the late 1990s, and again in 2001, left only four or five sites active on the Mississippi River mainstem (Clinton, Iowa; Grafton, Ill.; Thebes, Ill.; and St. Francisville, La.).
SOURCE: Alexander et al. (1997).

NASQAN sites from 1973 to 1995, and Figure 5-3 shows active NASQAN sites across the United States as of 2000.

The USGS implemented the National Water Quality Assessment (NAWQA) network in 1991, just as much of the NASQAN network was being eliminated (USGS, 2007). However, the NAWQA program does not represent a full replacement for NASQAN with regard to large rivers. Although NAWQA includes many Mississippi River tributaries, it includes no mainstem sites downstream of Lake Pepin. As a result, today only a few mainstem water quality sites remain in the USGS network downstream of Lake Pepin. These stations are at Clinton, Iowa; Grafton, Illinois; Thebes, Illinois; and St. Francisville, Louisiana. The NASQAN site on the Atchafalaya River at Melville, Louisiana, also could be included in this group because the Atchafalaya is the Mississippi River's primary distributary in the Mississippi's lower reaches (see Figure 5-3). Although some monitoring sites have been lost, a monitoring station at Belle Chasse, Louisiana, has come back online, and the USGS intends to bring another Atchafalaya River station online. The loss of monitoring sites of course represents the loss of future data from an individual site. However, a greater concern with the loss of water quality monitoring sta-

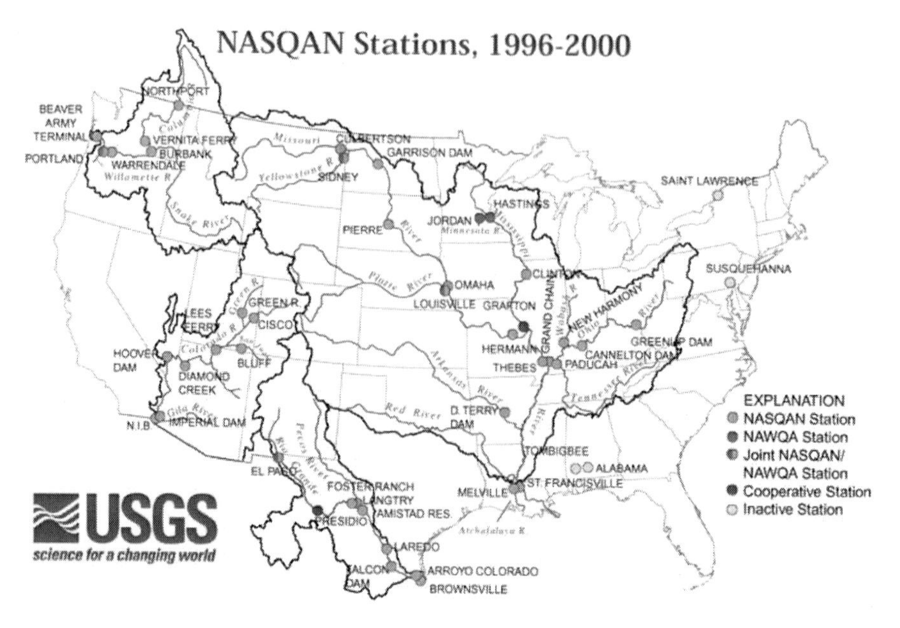

FIGURE 5-3 Active NASQAN stations as of 2000.
SOURCE: USGS (2006).

tions is that large-scale assessments, which could be useful in addressing regional or basinwide water management issues (e.g., hypoxia), cannot be replicated because more recent data from the same area have not been collected.

Comprehensive Mississippi River Assessments

Two widely cited water quality assessments that examine the Mississippi River at a regional or system-wide scale were published in 1995 and in 1999, and were headed by USGS scientists Robert Meade (1995) and by Donald Goolsby (Goolsby et al., 1999), respectively (these two reports are cited extensively in Chapter 2). In the 1995 report, the Meade team assessed water quality conditions along the Mississippi River mainstem from Minneapolis to New Orleans. They conducted longitudinal sampling on seven dates from 1987 to 1990 between St. Louis and New Orleans and on three additional dates from 1991 to 1992 between Minneapolis and New Orleans. Results from the 1995 Meade study were used by the Goolsby team as part of six reports that supported a Mississippi River assessment. Although these two USGS studies are a rich source of data in terms of both quality and quantity, they do not provide the coverage in space (many areas were unsampled) and time (these were snapshots or annual averages) needed to detect the frequency and duration of water quality standard violations for the Section 303(d) and Section 305(b) biennial assessments of the river required by the Clean Water Act (CWA). Furthermore, it is unlikely that these assessments can be repeated in the foreseeable future. Similarly, the 1999 USGS status and trends report for the upper Mississippi River, although a useful and creative synopsis of upper river ecology, is not Clean Water Act specific. That is, it is not aimed at determining if designated uses along the river are being met or assessing the frequency and duration of violations of water quality standards.

There have been other assessments of water quality along select portions of the river in addition to these studies. A 2002 report of water quality changes and conditions in the upper river near the Twin Cities is an excellent example (see Stoddard et al., 2002). However, no other studies have attempted to evaluate and characterize the entire river like the reports from the Meade and Goolsby teams.

The limited amount of water quality and river ecosystem data inhibits evaluations of lower Mississippi River water quality. The USGS conducts some sampling between Cairo and New Orleans, but this entails considerable difficulty, risk, and expense and therefore is very limited. Tennessee has conducted only modest data collection efforts, most of which are from the mainstem river directly downstream of Memphis. Arkansas, Kentucky, and Mississippi generally conduct minimal or no water quality sampling in the Mississippi River mainstem. Louisiana State University and the Louisiana Department of Environmental Quality have conducted more Mississippi River water quality sampling and have compiled their results into assessments.

There is a greater abundance of Mississippi River water quality data for the upper Mississippi River than for the lower river, due in part to efforts both of the federal-state EMP and LTRMP and of some upper river states. Through its NASQAN and NAWQA programs, the USGS has collected some water quality data for the Mississippi River, but these efforts have not been systematic and sustained, they have not been directed toward Clean Water Act objectives, and the resources allocated to these programs generally have declined over time. As the following section explains, the majority of water quality monitoring efforts along the Mississippi River aimed specifically at Clean Water Act directives have been conducted at the state level.

MONITORING ASSOCIATED WITH CLEAN WATER ACT OBJECTIVES

Monitoring and other techniques that determine whether water quality standards are met, including water quality and designated uses, are key steps toward achieving the Clean Water Act's "fishable and swimmable" objectives. Because states have the lead in implementing the Clean Water Act, monitoring and the design of monitoring programs are state, not federal, responsibilities. The Clean Water Act does not include any specific monitoring requirements, such as frequency of monitoring, parameters to be monitored, or locations for the siting of monitoring stations.

Water Quality Monitoring in an Interstate Setting

Several court decisions involving TMDL development have expressly refused to require the U.S. Environmental Protection Agency (EPA) to conduct water quality monitoring (*Sierra Club v. Hankinson*, 939 F. Supp. 865, 870 (N.D. Ga. 1996); *Ala Center for the Env't v. Reilly*, 796 F. Supp. 1374, 1380 (W.D. Wash. 1992), aff'd 20 F.3d981, 987 (9th Cir. 1994)). Others have found no legal mandate in the Clean Water Act for adequate state monitoring prior to EPA action to approve or disapprove a list of impaired waters that require TMDLs (*Friends of the Wild Swan, Inc. v. EPA*, 130 F. Supp. 2d 1184, 1193 (D. Mont. 1999); *Sierra Club v. EPA*, 162 F. Supp. 2d 406, 413 n.5 and 416 (D. Md. 2001)). At the same time, however, Clean Water Act Section 106(e)(1) conditions state receipt of federal grant funds for water pollution control programs on the EPA's finding that the state is monitoring the quality of its surface waters and compiling and analyzing the data obtained.

Water quality monitoring performed to meet Clean Water Act objectives has some recognized deficiencies, and some reports have confirmed the need to improve substantially the conduct of water quality monitoring, in both quantity and quality, as an essential basis for credible water quality improvement programs (NACEPT, 1998; GAO, 2000a; NRC, 2001; NAPA, 2002). For example, an EPA report commenting on state programs noted (USEPA, 2003c):

> States have taken very different approaches, within their resource limitations, to implement their monitoring programs. They have applied a range of monitoring and assessment approaches (e.g., water chemistry, sediment chemistry, biological monitoring) to varying degrees, both spatially and temporally, and at varying levels of sampling effort. It is not uncommon for the reported quality of a water body (i.e. attainment or nonattainment) to differ on either side of a state boundary. Although some differences can be attributed to differences in water quality standards, variations in data collection, assessment methods, and relative representativeness of the available data contribute more to differences in assessment findings. These differences adversely affect the credibility of environmental management programs.

Moreover, the discipline and practice of water quality monitoring does not always perfectly match CWA-related monitoring requirements. Water quality monitoring techniques and practices also are constantly being updated and improved (see Box 5-1).

Interstate waters such as the Mississippi River pose significant problems for the Clean Water Act framework. In addition to the size of such systems, political boundaries can create jurisdictional complications and make it difficult for individual states to commit resources to water quality monitoring in such waters. Moreover, given the Mississippi River's interstate nature, some states assume or assert that the monitoring and the condition of the river are exclusively federal responsibilities. A statement in the Mississippi Section 303(d) report for 2006 (MDEQ, 2006b) provides an example:

> The Mississippi Department of Environmental Quality (MDEQ) is not listing the Mississippi River on MDEQ's Mississippi 2006 § 303(d) list. In previous lists, the MDEQ included various segments of the river, but not based on data. Because any TMDL or delisting decision deals with multiple states and multiple EPA Regions, the MDEQ considers this a national issue. EPA Region 4 and Region 6 would jointly develop any TMDL for the Mississippi River.

At the national level, the EPA compiles the Section 305(b) assessments from each state into a national synthesis that is intended to indicate the condition of the nation's waters. In concept, at least, a similar approach could be used to assess the entire Mississippi River. There are, however, shortcomings with this approach, especially as it pertains to interstate rivers and to

BOX 5-1
What Is Water Quality Assessment?

Under the Clean Water Act, water quality assessments are technical reviews of physical-chemical, bacteriological, biological, and/or toxicological data and information to determine the quality of a state's surface water resources. Assessment begins with the assignment of appropriate designated uses for waterbodies and measurable water quality criteria that can be used to determine use attainment (*http://www.epa.gov/waterscience/standards/about/uses.htm*). The criteria, which may include biological, chemical, and physical measures, define the types of data to be collected and assessed. The EPA Office of Water has developed national indicators for surface waters and a conceptual framework for using environmental information in decision making (*http://www.epa.gov/waterscience/standards/about/crit.htm*).

In the more traditional approach to water quality assessment, monitoring data are compared to water quality criteria in order to make decisions on whether a waterbody is supporting (or not) its designated uses, such as aquatic life support, water contact recreation, and drinking water. This involves comparing criteria on a parameter-by-parameter basis. Basic limitations of this approach are (1) measurement of a set of individual physical, chemical, and biological parameters at numerous points in an aquatic system is expensive; (2) measurements often are available for only a few parameters; and (3) relating a set of parameter measurements to the health of an aquatic system is often difficult.

A newer, faster, and less expensive water quality assessment approach, which has emerged over the last two decades, is the use of rapid biological surveys, or rapid bioassessment protocols (RBPs). This approach is a response, in part, to dwindling resources available for monitoring efforts. It is also an attempt to evaluate biological conditions rapidly and the effects of water quality on those conditions in a particular system. In the RBP approach, surveys are conducted of aquatic macroinvertebrates, fish, or periphyton, and the presence or absence and relative abundance of species found is used to develop a numeric index that can be compared to a rating scale. This approach requires calibration to specific geographic area and, for the assessment of large rivers, is still in early stages of development.

Whichever assessment approach is used, a determination is made of whether the waterbody is fully supporting all of its uses; if not, the waterbody is considered impaired. The causes and sources of the impairment are then determined. Impaired waters are subject to further monitoring and are listed on the state's Impaired Waters List. The EPA has national guidance on assessing and listing impaired waters, known as the Consolidated Assessment and Listing Methodology (CALM), which generally undergoes revisions for each biennial reporting cycle.

an assessment at a regional or national scale. In particular, there is no scientifically defensible (i.e., statistical) basis for combining and extrapolating Section 305(b) assessments of individual waterbodies or reaches to make a quantitative statement about the extent, frequency, or fraction of compliance or noncompliance on a system-wide, regional, or national scale.

In a 2004 report, the Upper Mississippi River Basin Association noted shortcomings in using this approach to provide system-wide characterization of Mississippi River water quality:

1. Inconsistent designated uses of the river between bordering states;
2. Limited data to perform an assessment in most river sections or reaches;
3. Different judgments or differing water quality standards being applied by various states;
4. Different definitions of waterbodies being assessed (e.g., some reaches are defined and assessed differently among the states); and
5. Different time frames for the assessments (UMRBA, 2004).

Consequently, in terms of the Clean Water Act Section 305(b) framework, the Mississippi River is a patchwork of impaired and unimpaired sections, many of which overlap (Figures 5-4 and 5-5). Large sections of the river are evaluated in this process without supporting data, and some states ignore these interstate waters in their Section 305(b) process. As a result, opposite sides of the river sometimes have different impairment designations. Also, total miles of impaired and unimpaired reaches or sections can far exceed the total length of the river. Finally, states assess the condition of large portions of the river without reference to any actual water quality data. These difficulties are illustrated in reports from the Government Accountability Office (GAO, 2002) and the National Research Council (NRC, 2001) and are discussed in detail for the upper Mississippi River in a report from the Upper Mississippi River Basin Association (UMRBA, 2004). Coordination efforts among the upper Mississippi River states have alleviated some of these problems, but to date little has been done to coordinate such efforts along the entire river corridor.

One factor that encouraged the authorization and creation of the USGS National Water Quality Assessment program in the late 1980s and early 1990s was a recognition that large-scale compilation of state Section 305(b) assessments by the EPA could not provide a thorough, science-based assessment of the status or trend of conditions in the nation's waters. Unfortunately, the NAWQA design does not provide a complete national or regional assessment either, because its spatial coverage is limited and the design is not suitable for extrapolation of data to waterbody sections not studied. The EPA recently has increased its attention to this issue, possibly in the wake of several reports (e.g., GAO, 2000a, 2002) that identify deficiencies in state-level administration of CWA water quality standards and other aspects of the act.

One result of EPA efforts to improve water quality monitoring has been the National Wadable Streams Assessment (WSA), a program that aims to

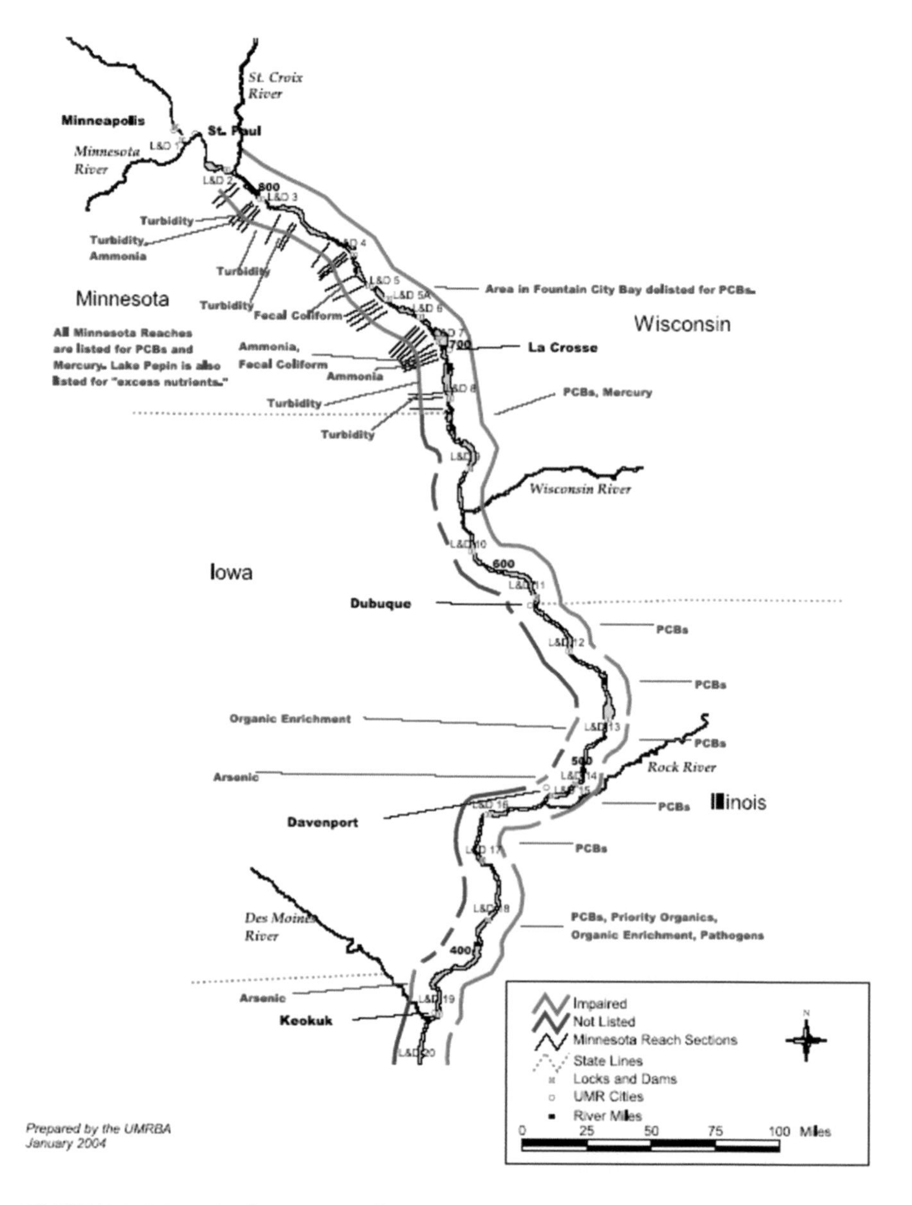

FIGURE 5-4 Impaired waters on the upper Mississippi River.
SOURCE: Reprinted, with permission, from UMRBA (2004). © 2004 by the Upper Mississippi River Basin Association.

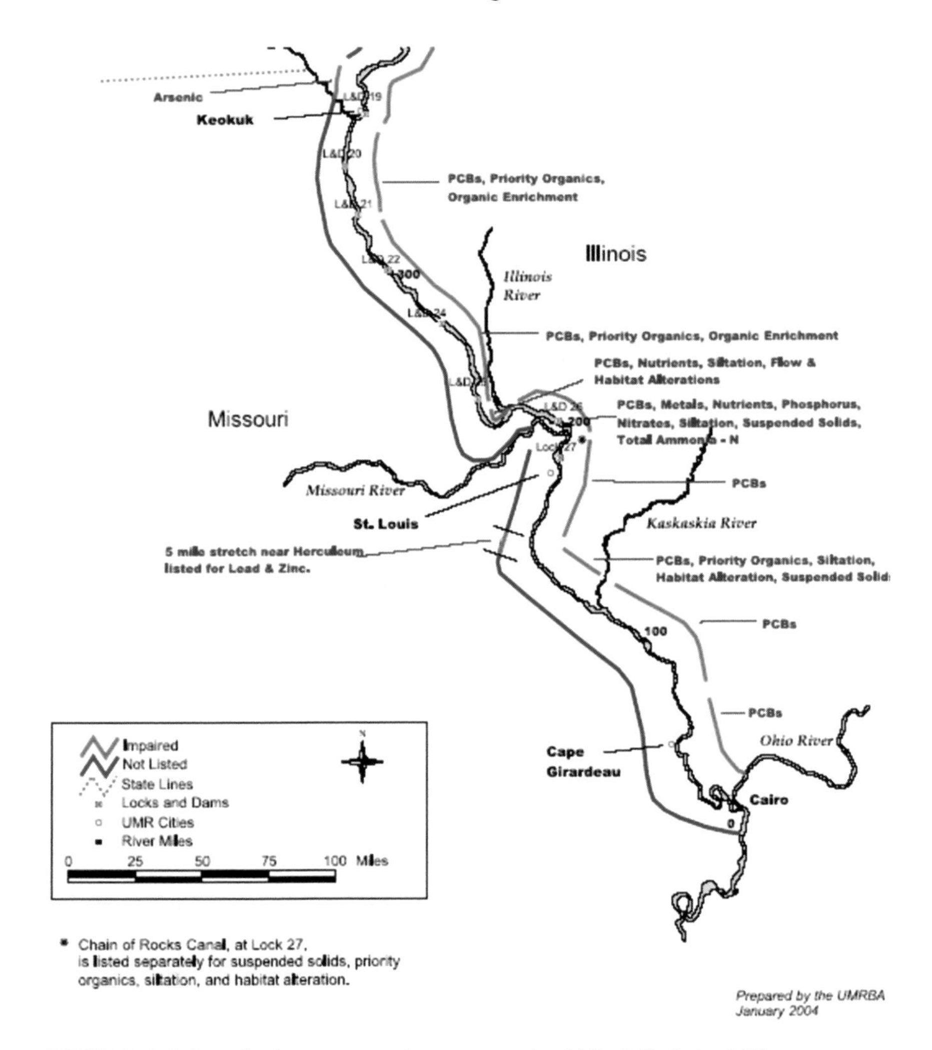

FIGURE 5-5 Impaired waters on the upper and middle Mississippi River. SOURCE: Reprinted, with permission, from UMRBA (2004). © 2004 by the Upper Mississippi River Basin Association.

minimize the sampling required to obtain a desired and statistically viable characterization. Large rivers are not yet included in the program. Also, the WSA approach focuses on biological indicators, for which many states have no numeric standards. However, the WSA does permit identification of stressors such as nutrients and sediments that can influence biota. Efforts to connect biotic indices with specific violations or remedial actions (such as TMDLs) on large rivers such as the Mississippi are in the early stages of development.

The EPA national water quality data repository known as STORET (storage and retrieval) contains water quality information for the Mississippi River and its tributaries (see EPA, 2007d). These data have been contributed by federal, state, and local agencies, volunteer and nonprofit groups, and others and are not based on a standard set of sampling or analytical protocols. Data quality is variable. The data included in STORET originally were obtained for a wide variety of purposes and generally do not derive from long-term, systematic monitoring efforts. The data in STORET may be useful for some kinds of evaluations, but to make use of STORET data for trend analysis at particular locations, it is often necessary to go back to the original data sources to understand sampling and analytical methods in order to assess the quality and comparability of data. The variability in type and quality of data in the STORET data repository system limits its usefulness for water quality assessments.

State Monitoring Programs

The 10 states along the Mississippi River mainstem conduct limited monitoring in the river (Table 5-1). Because of longitudinal and subbasin

TABLE 5-1 Status of Water Quality Monitoring in Mississippi River Mainstem States

State	Constituent Monitoring?	Biological, Wildlife, and Habitat Monitoring?
Minnesota[a]	Yes, pools 1-4	Pools 1-4
Wisconsin[a]	Yes, pools 4-12	Irregular
Iowa	None	Irregular
Illinois[b]	Pools 13-26 and middle river (below St. Louis)	Irregular
Missouri[c]	Limited—irregular	Irregular
Kentucky[d]	Limited	Not main channel
Tennessee[e]	Memphis area	Irregular
Arkansas	None	None
Mississippi	None	Irregular—none
Louisiana[f]	Significant	Yes

[a]Minnesota-Wisconsin—substantial monitoring documented in Sullivan et al. (2002).
[b]Monitoring increased dramatically in 1999 with mainstem sampling conducted at about 50-mile intervals.
[c]Little to no monitoring, a significant amount of sampling conducted by (or for) USGS.
[d]Has almost no border on the mainstem; conducts little or no sampling.
[e]Almost no monitoring, limited to immediate vicinity of Memphis.
[f]Significant programs (see Box 5-3). The last several hundred miles of the Mississippi River flow entirely within Louisiana so it is a water of state interest and responsibility.

differences in the character of the Mississippi River, the water quality issues of greatest priority differ among Mississippi River reaches and over time. State monitoring efforts in the Mississippi River typically have been directed to specific issues or crises. Maintenance and updating of fish advisories (Box 5-2) is a common motivator of monitoring efforts. For monitoring conducted to address water quality questions other than fish contamination, when the questions of local program sponsors are answered (or become less urgent), the programs are often scaled back, discontinued, or refocused. This tends to produce short-term, focused sampling designs yielding data that cannot readily be combined across differing locations or time periods to produce a consistent, long-term, large-scale view of the system.

For example, Minnesota has a statewide surface water monitoring system organized on a basin-watershed basis, with two of the ten major drainage basins related to the Mississippi River, but there is limited direct monitoring in the Mississippi River. The statewide system includes routine chemical monitoring at 80 fixed stations throughout the state, with 3 sites on the Mississippi River border with Wisconsin (UMRBA, 2004). Minnesota also conducts biological monitoring at 55 randomly selected locations in each of the state's 10 main hydrologic basins. A separate fish contaminant monitoring program is conducted for updating fish consumption advisories and aiding Section 303(d) impairment assessments.

Wisconsin conducts baseline monitoring of nonwadable streams at 180 sites statewide, but no sites in the Mississippi River are included in this program. The state does conduct routine water quality monitoring at three lock-and-dam sites on the river, although this is not part of the baseline monitoring program (UMRBA, 2004). Wisconsin also performs regular monitoring of fish contaminants.

Illinois has maintained a limited monitoring program of the mainstem of the Mississippi River for more than 30 years. In 1999 this program was expanded to provide 11 sampling sites—8 at locks and dams and 3 open river locations. The 11 mainstem stations are sampled quarterly for basic chemical parameters and other parameters indicative of river health. In addition, seven sites are tested routinely for chlorophyll and two sites are tested for pesticides. In 2001, macroinvertebrate sampling was added at a number of the sites. The stations are spaced at approximately 50-mile intervals, with the eight stations above St. Louis located at locks and dams and the three stations below St. Louis sampled from boat ramps (UMRBA, 2004). Illinois is the only state below its border with Wisconsin that conducts a structured monitoring program for the Mississippi River.

Iowa does not conduct monitoring on the Mississippi River, and although Iowa significantly expanded its statewide Ambient Monitoring Program 1999, it still does not include any sites on the Mississippi River. Iowa therefore relies entirely on other data sources for assessing the upper

BOX 5-2
Mississippi River Monitoring and Fish Consumption Advisories

A primary motivator of state-conducted monitoring of the Mississippi River is the protection of fish resources and maintenance of up-to-date fish consumption advisories. In fact, the issue of fish contamination is one of the greatest concerns of sport and commercial fishermen and the general public along the upper Mississippi River. Many commercial and recreational anglers depend heavily on Mississippi River fishery resources, and many regional and local community economies are supported by recreational use and river-related tourism.

Fisheries are jeopardized when toxins contaminate fish by direct exposure to water or sediments or through the food chain. Some of these contaminants are legacy materials (e.g., PCBs [polychlorinated biphenyls], DDT [dichlorodiphenyltrichloroethane]) and some derive from current practices (e.g., mercury, dioxin, lead). These toxins can accumulate in fish tissue over time and reach concentrations that pose a risk to human health. Concentrations of toxic substances in fish tissue can be much higher than those found in the water.

States along the river monitor various fish species and use different approaches for assessing health risks. The states publish Fish Consumption Advisories (FCAs) that recommend limits on the consumption of fish, and they decide if a river segment should be listed as impaired under the Clean Water Act because of this contamination. Along some segments of the river, bordering states have issued different FCAs and have categorized the impairment of the river section differently. This can lead to public confusion about the risks from fish caught in the river and can have economic and regulatory implications for point source dischargers to the river (FTN Associates, Ltd. and Wenck Associates, Inc., 2005).

Evaluations of fish tissue quality differ from traditional water quality assessment, which involves measurement of a particular water quality parameter and comparing it to a criterion. Fish tissue analysis provides an aggregate measure of aquatic organism exposure to a range of contaminants. Such analyses are used in water quality impairment assessments and also support public health protection through issuance of FCAs. The FCA process starts with collection and analysis of fish tissue, proceeds to an evaluation of the risk to human health, and then estimates what consumption limit (e.g., frequency and amount) should be recommended for specific users (e.g., children, pregnant women) of specific fish types (e.g., fish species, size, body portions) taken from specific areas. If fish contaminants exceed a certain level or a FCA is issued for a waterbody, the river segment may be added to the Clean Water Act Section 303(d) list of impaired waterbodies.

The states, District of Columbia, U.S. territories, tribes, and local governments have primary responsibilities for protecting their residents from the potential health risks from eating contaminated fish caught in local waters. The states have developed their own fish advisory programs over the years, and there are variations among them in terms of extent of monitoring, frequency of sampling, decisions made regarding advisories, and so on. EPA plays a role in providing a National Listing of Fish Advisories database. This is an annual compendium of information on locally issued fish advisories and safe eating guidelines that is provided to EPA by the states and other bodies. EPA has compiled and made this information available since 1993 (available online at *http://www.epa.gov/waterscience/fish/advisories/2006/index.html#basic*).

Mississippi River in its Section 305(b) assessments and Section 303(d) lists. Differing combinations of data sources are used to evaluate each of Iowa's 14 upper Mississippi River reaches (UMRBA, 2004).

With the possible exception of Louisiana, monitoring downstream of the Ohio River confluence that is related to Clean Water Act assessment, enforcement, and restoration is less active than in the upper Mississippi River states. In general, the lower Mississippi River states consider Mississippi River water quality to be the responsibility of others and give it low priority for monitoring funds. For example, Mississippi and Arkansas provide an example of limited involvement in Mississippi River monitoring, because they no longer assess the Mississippi mainstem as part of their Section 305(b) process.

STATUS OF AND PROSPECTS FOR MISSISSIPPI RIVER MONITORING

Current Efforts

The status of monitoring on the Mississippi River to obtain data relevant to Clean Water Act assessment and enforcement presents a mixed picture. Assessment of water quality and habitat for the Clean Water Act has been done relatively well on the upper river, but even there, there are limitations within the data gathered to date. Furthermore, levels of commitment of the 10 Mississippi River states to river monitoring are varied and may change in the future. Data collected often are not readily comparable (Box 5-3). Federal monitoring programs on the Mississippi River are focused on fish and wildlife populations, habitat conditions, and mass transport of nutrients and sediments. These programs are not designed to be part of CWA-related monitoring (e.g., verifying whether a given state's designated uses are being attained).

The limitations of federal monitoring programs on the Mississippi River are illustrated within the upper Mississippi River LTRMP. This program has the primary purpose of monitoring biotic conditions and habitat at a system-wide, multiyear scale. The water quality data collected by this program are a primary source of information for substantial portions of the upper river, and although it has been useful in Clean Water Act assessments (e.g., by Minnesota), the LTRMP is focused on habitat conditions and is not intended to track compliance with water quality standards. Thus, the program does not monitor a host of pollutants that have numeric standards and are priority pollutants of regulatory interest under the Clean Water Act, nor does LTRMP monitoring lend itself to the detection of short-term, acute conditions (e.g., violations of water quality standards) at specific locations for specific durations or frequencies. Further, this program, like many other

BOX 5-3
Consistency of Water Quality Data

The ability to combine or compare data from different sources is an important issue with no easy solution. Data can differ not only because of differing methods or equipment used to collect and analyze water samples, but also because of differences in sampling design (i.e., what basic aspects of the system are represented in the data; see Soballe, 1998). For example, data that are collected during midday sampling must be adjusted before they can be combined or compared with data from a program that samples only at night or during pre-dawn hours. Such an adjustment may not be possible. Likewise, data that are collected only to represent high-flow storm conditions in one location may not be easily combined or compared with data that sample end-of-pipe or low-flow conditions in another. An approach often used in stream sampling programs is to collect a "flow-weighted" sample in which a single sample is generated for chemical analysis by adding water to a single container for several hours or several days in proportion to the river's surface elevation or flow. Such a sample is useful for calculating mass transport, particularly during a single rainstorm or flood; however, results of this flow integration are not readily comparable to those produced by sampling at regular, longer-term intervals (weeks or months) to detect extremes or to estimate average conditions. There is no single standard method that can be applied to all sampling to meet all information needs.

federally sponsored efforts, has been reduced since the late 1990s and has been forced to focus more closely on its primary mission of tracking the status of biota and habitat in specific study areas. Although the LTRMP has collected data from thousands of locations along the river for more than 15 years, these efforts have tended to be seasonal and limited to five river reaches. There has been no mechanism to extrapolate these data to intervening portions of the river or to other periods of time. Data collected by the program clearly have value for improved understanding of Mississippi River aquatic ecosystems (see, for example, USGS, 1999), but they have limited utility regarding CWA-related assessment of the entire system.

The seemingly low level of Clean Water Act-related monitoring on portions of the Mississippi River is not unique or even unusual. For example, the GAO reported that as of 1996, states assessed only 19 percent of their rivers and streams (GAO, 2000a). The GAO also noted that states tend to focus monitoring on those waters with suspected pollution problems in order to direct scarce resources to areas that could pose the greatest risk (GAO, 2000a). Because of the dilution capacity of the Mississippi River, the difficulty of large-river water quality monitoring, and the absence of sole responsibility of individual states for its quality, states have given Mississippi River monitoring low priority.

A wide range of water quality and ecosystem monitoring efforts have been and continue to be conducted along the Mississippi River. These efforts are quite variable in spatial and temporal implementation, are not well coordinated, and for the most part are not designed for Clean Water Act assessment purposes. Better coordination and a shared sense of purpose and value of monitoring information among the mainstem river states are needed for more effective and useful system-wide monitoring.

The Value and Importance of Monitoring

Monitoring of Mississippi River water quality has not been performed in a system-wide manner for extended periods (e.g., decades) and at intervals of time (e.g., monthly) or space (in every major reach) that would support rigorous assessment of water quality and ecology for the river. As this chapter has discussed, there are considerable challenges to conducting this type of extensive monitoring: large-river sampling methods and instrumentation need to be standardized; states and federal agencies must compare and cooperate on sampling and monitoring strategies to make the most of their expenditures and prevent duplication of ongoing efforts; the resources required for extensive and sustained monitoring can be considerable; and there are practical challenges to monitoring, especially in the often dangerous lower Mississippi River.

Despite the costs and analytical and logistical challenges involved in creating such a program, there are also costs in not having a systematic monitoring program for the entire Mississippi River and into the Gulf of Mexico. The nation's rivers, including the Mississippi, have realized improvements in some aspects of water quality as a result of the Clean Water Act. Many of those improvements have been achieved through reductions in point source discharges of pollutants.

Water quality issues and problems of primary concern along the river today are different than in the early 1970s when the Clean Water Act was enacted and consist primarily of nonpoint pollutant loads from agricultural, urban, and suburban activities. The framework within the Clean Water Act for addressing nonpoint source pollutants relies more strongly on scientific data, monitoring, and modeling of water quality than on an end-of-pipe approach to treating point source pollution (Box 5-4 discusses the role of modeling in water quality assessments). Rather than focusing on reducing discharge from individual sites, contemporary programs for achieving water quality improvements in the Mississippi River and the Gulf of Mexico must encompass pollutant inputs from across the entire watershed. They must also monitor water quality conditions for the river as a whole, not just at

BOX 5-4
Role of Modeling in Water Quality Assessment

Although acquisition and analysis of monitoring data is the approach preferred by the EPA for identifying impaired waters, modeling can have an important supplementary role. Integrated monitoring and modeling can often provide better information than monitoring alone for the same total cost (NRC, 2001). For example, Section 303(d) and related guidance from EPA recommend focusing efforts on waterbodies or segments that are suspected of violating water quality standards. Such targeted monitoring represents the use of available information regarding water quality impairments to guide monitoring toward particular sites. A potentially valuable use of modeling in relation to Section 303(d) listings would be to formalize the use of available information on impairment probability in monitoring system design. Limited monitoring resources could be focused on sites where impairment is most uncertain, thus improving the efficiency of monitoring.

points near specific sources of effluent. Today, water quality improvements rely more heavily on a science- and data-intensive approach to understanding the linkages between activities that generate pollutant loads and their ultimate impacts on waterbodies. Without comprehensive monitoring of a river system, it is difficult to understand trends in water quality conditions, to realize the impacts of watershed-focused programs designed to reduce nutrient and sediment loads, and to determine whether designated uses are being achieved.

Beyond limited amounts of data, another challenge to system-wide assessment is that some of the data collected by the many state and federal monitoring programs have fundamental differences in their underlying purposes and designs (Box 5-4). When monitoring program details are compared, it is often discovered that data from different sources cannot be combined in a meaningful way. Thus, the ability to compare data over large scales of time and space is further restricted. The situation is created, in part, by the scales of time and space required for adequate research and monitoring and by the specific issues the monitoring system is designed to address. These scales are dictated by natural scales of the system and the questions being addressed (Soballe, 1998), and the questions and issues have seldom been the same across multiple monitoring programs.

For the Mississippi River, the lack of a coordinated water quality data gathering program and of a centralized water quality information system hinders effective implementation of the Clean Water Act and acts as a barrier to maintaining or improving water quality along the river and in the Gulf of Mexico. The EPA should take the lead in establishing such a program. In doing so, it should work closely with the 10 Mississippi River mainstem states and with federal agencies with relevant expertise and data, such as the Corps of Engineers, the USGS, and the National Oceanic and Atmospheric Administration (NOAA). Part of this effort should focus on collecting data necessary to develop numeric water quality standards for nutrients in the Mississippi River and the Gulf of Mexico.

Emerging Monitoring Challenges

Some emerging developments in aquatic system monitoring pose particular challenges for implementation in large systems such as the Mississippi River. There is increasing interest in biological monitoring because of the direct link to ecosystem health and the potential to evaluate the aggregate impact of water pollution. Techniques and biocriteria have been developed for smaller streams, but neither have been established yet for large rivers. There also have been advances in tracking sources of sediment inputs to streams.

Biomonitoring

In many Clean Water Act assessments, the condition of a waterbody with respect to supporting a designated aquatic life use is evaluated primarily through stream biological community assessments. Biomonitoring of resident biota can often be conducted more quickly and less expensively than monitoring of physical-chemical water quality parameters. Bioassessment protocols (e.g., rapid bioassessment protocols; see Box 5-5) could fill some data gaps with regard to the Mississippi River CWA-related assessments, but this approach has been limited to date to wadable streams. In addition, meaningful biocriteria (numeric measures of desirable fish populations, etc.) for large rivers have not been established, nor have means been developed to readily collect the necessary data for sound bioassessments of large rivers.

Impairments for human contact or consumption can also be assessed using fish tissue analyses and evaluations of raw (intake) water monitored by water purveyors. These are used in some reaches of the Mississippi mainstem, but their application seems to be less than consistent (UMRBA, 2004). Recreational use impairments are often based on bacteriological data, such as fecal coliform counts, and these are commonly used, at least in the upper river.

Sediment Monitoring

Sediment concentration and transport are crucial water quality and river ecology issues along the entire Mississippi River, but systematic monitoring

BOX 5-5
Rapid Bioassessment Protocols

Biological surveys and other direct measurements of the resident biota in surface waters are often used to determine whether a surface waterbody is meeting a designated aquatic life use. The Rapid Bioassessment Protocols (Barbour et al., 1999) were developed in the 1980s and 1990s by various states and compiled by the EPA in a guidance document. The guidance includes protocols for three types of aquatic life—periphyton, benthic macroinvertebrates, and fish—as well as for habitat assessment. These protocols have all been tested in streams and wadable rivers in various parts of the United States. They have been used as rapid, inexpensive means of water quality assessment and have been utilized extensively by states in development of Section 305(b) water quality inventories. Bioassessment has also been used in Section 303(d) impairment assessments. Effects of excess nutrients, sediments, and other pollutant classes can be readily identified.

Bioassessment protocols that are practicable and can be linked unequivocally and quantitatively to the functional health (or biotic integrity) of the large river are still under development. A limitation of the use of bioassessments for evaluating conditions in large rivers such as the Mississippi River is the difficulty in linking biological metrics unambiguously to specific causal factors. Thus, it currently is not possible to initiate specific remedial action or management based on the numerical value of bioassessment indices alone. However, these indices can be valuable for identifying the need for a more detailed evaluation of conditions in impaired locations.

of these important variables poses analytical and conceptual challenges. Standard, widely accepted approaches to assessing sediment dynamics (i.e., deposition and resuspension) have not been developed and accurate measurements of sediment dynamics over long time periods (years) and large spatial scales (tens to hundreds of kilometers) are difficult to obtain. For example, reports on Mississippi River water quality and ecological integrity often note sediment, "siltation," and turbidity as priority concerns in the upper river (UMRBA, 2004; Headwaters Group, 2005). These various terms are interrelated and, although sometimes used interchangeably, do not have the same meaning. Monitoring data for any one of these characteristics are not particularly informative about the others. Turbidity, for example, is governed by the size, composition, and concentration of suspended particles in the water. It can be viewed as a short-term, near-field property because the particles that create this phenomenon may change rapidly (minutes) over short distances (meters) in the river. Monitoring that does not capture these short-term, near-field variations may not reveal the extremes of turbidity to which river biota are exposed. Smaller particles (fine silts and clays) have the greatest influence on turbidity, but coarser particles (sand) usually dominate the process of sedimentation. In contrast to turbidity, sedimentation is a longer-term processes (years to centuries), and assessment of this phenomenon depends on the scales of space and time used in the measurement. Moreover, the data used to study these processes on the river generally are sparse (see Box 5-6).

As Figure 5-6 illustrates, an area that appears to be accumulating sediment for several years or decades may be deeply scoured by occasional large floods and therefore be in dynamic balance over the longer term. Likewise, one portion of a river reach may be accumulating sediment, while an adjacent zone is being scoured, so that on a larger spatial scale, the total reach appears to be in balance. However, this balance may be only temporary and extend over a few years or a few decades.

These complications and variations over time regarding sediment transport and loadings are illustrative of the larger challenges that attend accurate and consistent monitoring of water quality variables and provide background for the following conclusions regarding federal and state water quality monitoring programs along the Mississippi River.

BOX 5-6
Sediment Transport and Deposition: A Monitoring Challenge

A study of one of the upper Mississippi's tributary streams—Coon Creek, in Wisconsin—demonstrates some of the complex patterns of sediment transport and deposition in a single stream, how those patterns may change over time (Figure 5-6), and the kind of monitoring needed to study long-term sediment transport and deposition. Research in Coon Creek has shown that sediment yield varies depending on where it is measured within a basin. Despite a significant decrease of sediment flux within the basin caused by improved land management practices, sediment yield from Coon Creek to the Mississippi River has held fairly constant (at least according to available data). As indicated, this continued flow of sediment is coming from upstream channels and banks.

There is presently only one sediment measuring station in this entire region for tributaries to the Mississippi River. However, measurements on the main river downstream at Dubuque, Iowa, indicate that sediment transport is presently only about half the rate existing in the 1940s (Pannell, 1999). How can this apparent disparity be explained? Are either or both measures wrong? There simply are not enough sediment measuring stations to know.

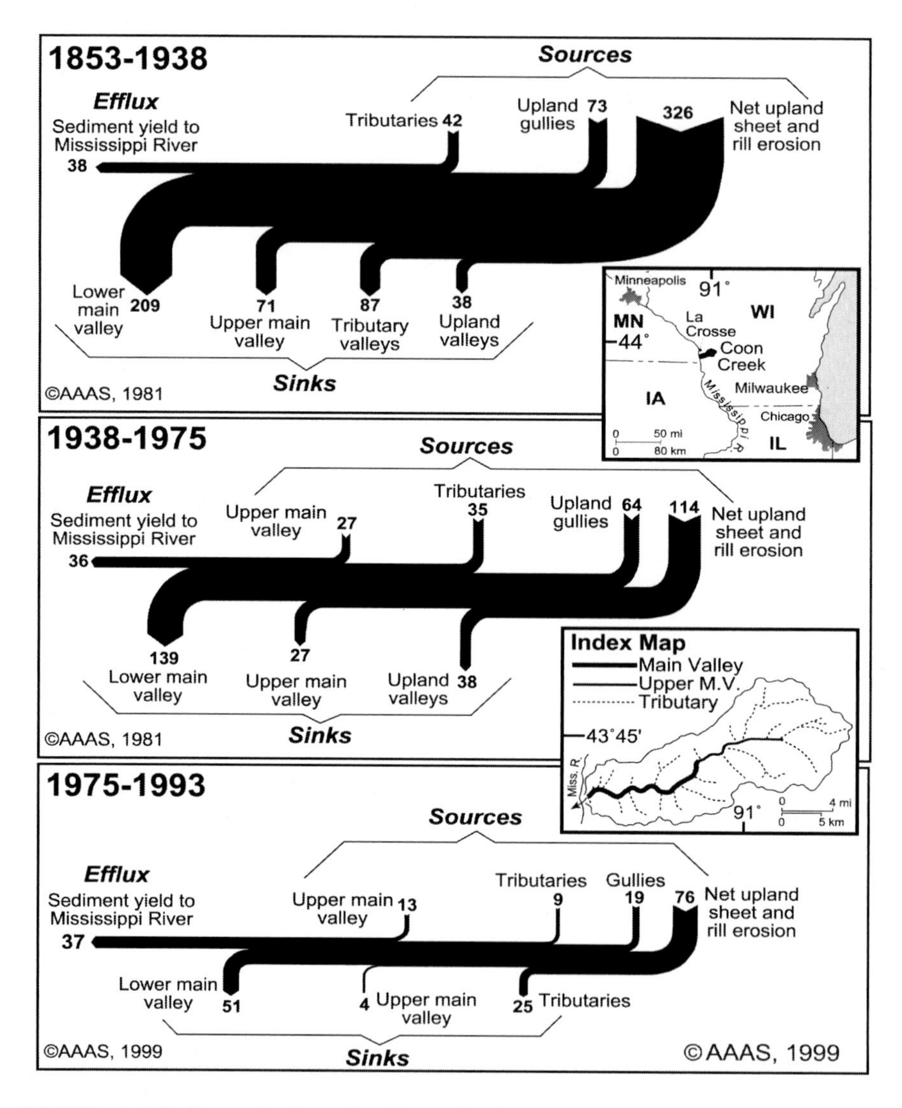

FIGURE 5-6 Sediment budgets for Coon Creek, Wisconsin, 1853-1993. Numbers are annual averages for the periods in thousand tons per year. All values are direct measurements except "net upland sheet and rill erosion," which is the sum of all sinks and the efflux (sediment yield to the Mississippi River) minus measured sources. The lower main valley and tributaries are sediment sinks, whereas the upper main valley is a sediment source.
SOURCE: Reprinted, with permission, from Trimble (1999). © 1999 by the American Association for the Advancement of Science.

SUMMARY

Restoration and maintenance of water quality and related ecosystem conditions in the Mississippi River require understanding of current system conditions and trends over time. Monitoring data are necessary for the assessment and planning required under the Clean Water Act to maintain and improve water quality. However, large rivers such as the Mississippi are difficult to monitor consistently and comprehensively for water quality and biota. High-quality monitoring programs require expensive and rugged equipment, specially trained personnel, and more time in the field than is needed to monitor streams and small rivers. Further exacerbating the challenge of assessment of the Mississippi River is the fact that water quality and habitat differ across the river's many subbasins. Along the Mississippi River, there are large longitudinal gradients in water quality, geomorphology, and biota.

There is no consistency in the amount and quality of water quality data available for the length of the mainstem Mississippi River. Some areas in the upper river have been relatively well monitored and there is a large amount of water quality data. At the federal level, these efforts primarily are represented by the EMP and LTRMP. Data from the LTRMP could be useful in a supplementary role in Clean Water Act assessments, but the LTRMP is focused on habitat conditions and is not intended to track compliance with water quality regulations. The USGS also has collected some Mississippi River water quality data via its NASQAN and NAWQA programs, but these efforts have not been systematic and sustained, they have not been directed toward Clean Water Act objectives, and the resources allocated to the programs have generally declined over time.

On the upper river, Minnesota, Illinois, and Wisconsin have promoted the most extensive Mississippi River programs at the state level, although the resources devoted to these programs have varied over time. In the lower river states, there are fewer data and there have been far fewer monitoring initiatives. Tennessee has conducted only modest data collection efforts, most of which are on the mainstem river directly downstream of Memphis. Arkansas, Kentucky, and Mississippi generally conduct minimal or no water quality sampling in the Mississippi River mainstem. Louisiana has conducted more Mississippi River water quality sampling and has conducted some assessments with the results. Some of these upstream-downstream differences are driven by different values and uses of the respective portions of the river; the physical difficulties and hazards posed by monitoring in the large lower Mississippi River also are factors.

Water quality monitoring along the Mississippi River mainstem is inconsistent over both space and time. The extent to which Mississippi River mainstem states monitor water quality in the river varies considerably, and these efforts lack coordination. States along the river have assigned different designated uses to the same river segments; they use different judgments and methods in their assessments; and there is no standard for the time frame or frequency of water quality monitoring. Mississippi River monitoring programs conducted by the USGS and the Corps of Engineers have diminished over time in many places, although the USGS is increasing monitoring capabilities and the number of stations in some areas (e.g., Atchafalaya River). Generally speaking, the extent and quality of biological, physical, and chemical data along the river generally do not support thorough CWA-related assessments. **The lack of a centralized Mississippi River water quality information system and data gathering program hinders effective application of the Clean Water Act and acts as a barrier to maintaining and improving water quality along the Mississippi River and into the northern Gulf of Mexico.**

States along the mainstem Mississippi River, together with the federal government, need to coordinate better with respect to planning monitoring activities and sharing the data that result. In a climate of ever-decreasing resources for monitoring, all federal and state agencies involved in monitoring the Mississippi River mainstem should cooperate and coordinate their efforts to the greatest extent possible. The Mississippi River clearly is of federal interest because of the many states in the river basin, the river's prominent role in supporting interstate commerce, and its hydrologic and ecological systems that extend across several states and into the Gulf of Mexico. The federal government should take the lead in ensuring adequate water quality monitoring, a cornerstone of effective Clean Water Act implementation along the Mississippi River and into the Gulf of Mexico.

There is a clear need for federal leadership in system-wide monitoring of the Mississippi River. The EPA should take the lead in establishing a water quality data sharing system for the length of the Mississippi River. This would include establishing coordinated monitoring designs and developing mechanisms (hardware, software, and protocols) necessary for efficient data sharing among monitoring and resource agencies and Section 305(b) and Section 303(d) assessment teams. It also would entail ensuring consistency in river monitoring in terms of parameters measured, units and methods employed, and siting of monitoring stations along the length of the river. The EPA should draw on the considerable expertise and data held by the U.S. Army Corps of Engineers and the USGS, as well as NOAA and the water-related data for the northern Gulf of Mexico that it collects and maintains. The EPA should work closely with Mississippi River states in establishing this plan and system. A priority for EPA in this regard should be to coordinate with the states to ensure the collection of data necessary to develop numeric water quality standards for nutrients in the Mississippi River and the Gulf of Mexico.

6

Agricultural Practices and Mississippi River Water Quality

Agricultural land uses and practices are of central importance to nutrient and sediment loads into the Mississippi River and the Gulf of Mexico and merit discussion in a broad review of Mississippi River water quality issues. As explained in Chapter 2, agriculture is the predominant land use across the Mississippi River basin, and agriculture is central to both nonpoint source pollution issues and water quality restoration strategies.

The farming of row crops, such as corn and soybeans, in the basin has increased over time, and there has been a corresponding increase in mean nitrate concentration in runoff across the basin. The soil system has lost nitrogen as farmers have plowed under prairie grasses and exposed the soil. Moreover, since World War II, farmers have increasingly used nitrogen fertilizers to support the growth of crops. Today, phosphorus and nitrogen loadings to the Mississippi River are predominately from agriculture, with loadings from municipal and industrial point sources representing only a small fraction of that contribution (Goolsby et al., 1999, and Figure 2-11).

As Chapter 2 emphasizes, the primary nonpoint pollution concerns in the Mississippi River basin are nutrients, which derive largely from fertilizers applied to crops, and sediments, which derive largely from soil erosion and are related to tillage. Agricultural practices therefore are key factors in efforts to address both of these critical pollutants in the Mississippi River and the Gulf of Mexico.

As this chapter discusses, agricultural practices and policies involve a trade-off between protecting water quality and related environmental services, on the one hand, and efforts to increase production of food, fiber, and most recently, bioenergy, on the other. Historically, agricultural policies and programs have emphasized agricultural commodity production. More recently, Congress and the U.S. Department of Agriculture (USDA) have created and implemented agricultural programs that facilitate conservation of land and water resources, but these programs generally have received far less emphasis than crop production incentives. Nevertheless, the balance between these two goals has been shifting. Many farmers today across the river basin are seeking ways to improve farming and production efficiencies, while at the same time seeking to increase environmental benefits. These latter benefits can also be viewed as a type of "commodity," albeit a nontraditional one.

This chapter discusses agricultural production and conservation programs, including strategies for reducing sediment and nutrient loadings within the context of the Clean Water Act and through federal and state agriculture-related initiatives. It identifies and describes existent and emerging regulatory, incentive-based, and market-based approaches for reducing nonpoint inputs. It also provides recommendations for ways in which the states, the USDA, and the Environmental Protection Agency (EPA) might strengthen cooperative efforts to improve water quality through agricultural programs and actions.

TENSIONS BETWEEN AGRICULTURAL PRODUCTION AND WATER QUALITY

The Farm Bill

The Agricultural Adjustment Act of 1933 established a time-honored tradition in American agriculture: the notion that it is necessary to control the supply of agricultural commodities in order for farmers to receive a fair price for their goods (Cain and Lovejoy, 2004). The act pursued this goal by setting price supports, or parity prices, to guarantee that prices did not fall below a set level. This price support was available to producers who participated in voluntary production reduction programs, such as acreage set-asides. Early farm bills defined a pattern of government involvement that still holds today: voluntary participation based on economic incentives through income or price support and payments for specific actions. Today, most government payments subsidize producers of commodity program crops such as corn, wheat, soybeans, cotton, rice, and peanuts.

Commodity payments and price supports can lead to more extensive and intensive production than would be the case if there were none, because these mechanisms give farmers an economic incentive to expand actual and potential crop production. However, these rational responses to the Farm

Bill's economic incentives carried with them attendant impacts on land use, runoff, and water quality. Historically, such payments also linked a specific land base that produced the commodity to the payments. For example, the 1996 Farm Bill changed direction from commodity acreage-based payments to farm-based payments (Schertz and Doering, 1999). However, these general income payments under the 1996 act still helped farmers maintain production levels during a period of relatively low commodity prices in the late 1990s.

The 1933 Farm Bill, and the subsequent 70 years of Farm Bills and other agricultural programs, have had a tremendous influence on Mississippi River basin land uses, crop types, farmer attitudes and preferences, and the structure of the agricultural sector; in turn, they have greatly affected runoff patterns and water quality across the basin and in the Mississippi River and the Gulf of Mexico. Nevertheless, Farm Bills also have contained provisions encouraging conservation practices, and the 2002 Farm Bill included an unprecedented expansion in federal support for farmers for conservation activities by introducing the Conservation Security Program and continuing the Environmental Quality Incentives Program.

Impacts of Commodity Programs on Production and Conservation

Commodity program benefits have led to modest increases in acreage of program crops (Young and Westcott, 2000). Historically, when price supports or market prices change to alter a long-term market price ratio (for example, between corn and soybeans), there are acreage shifts in the Mississippi River basin to more profitable crops. Such crop shifts may cause more or less sediment or nutrient loss in the basin. Technological changes and advances may also affect crop mixes and land uses. In low-moisture grassland and prairie environments, for example, the development of herbicide-resistant soybeans, combined with no-till planting technology, allow such lands to be planted in soybeans. Favorable prices or government support programs may also be required to encourage soybean planting in these areas.

Commodity program payments sometimes compete with incentives for farmer participation in voluntary land and water conservation programs. It is clear that higher commodity revenue, whether through the market or through price supports, provided through government programs means that farmers will require increased incentive payments to engage in farm-level conservation and water quality-enhancing activities (Moore, 2002). Higher crop prices also tend to increase land values, making land retirement-based programs more expensive.

Many other long-term and structural effects of commodity program benefits influence crop types and levels of production:

- *Wealth shift.* Payments increase the overall wealth of farmers, increasing investments in productive assets and enhancing production. Increases in land values are an important component of this effect (Young and Westcott, 2000).
- *Greater access to credit.* Lenders are more willing to lend money based on the more stable stream of income that commodity payments and insurance provide.
- *Risk aversion.* Reduction in risk also encourages producers to maintain or increase production levels from where they might be otherwise (Chavas and Holt, 1990).
- *Expectations about future programs.* Because the tradition has been one of program payments based on past planted acres, producers can be reluctant to give up the production of program crops on a given tract of land. In addition, there is an expectation of continuing government support payments into the indefinite future.

Commodity program benefits also have important effects on marginal agricultural lands. Crop insurance, for example, disproportionately keeps in production low-productivity land and some environmentally sensitive lands such as those with highly erodible soils. Also, the land retained in cultivation because of crop subsidy increases includes a higher proportion of lower-quality land than the national average for cultivated cropland. Such low-productivity land leaches higher amounts of nitrogen and adds greater amounts of phosphorus to surface waters (Lubowski et al., 2006). Commodity programs thus provide incentives for production that may work against farmer participation in voluntary land and water conservation programs. The federal government, through the USDA, has created programs that aim to balance incentives for production with incentives for conservation and environmental quality improvement.

FEDERAL AGRICULTURAL PROGRAMS FOR RESOURCE CONSERVATION

To encourage land and water quality conservation practices, the USDA sponsors several programs that provide incentives for voluntary participation. The largest of these land and water conservation programs are the Conservation Reserve Program (CRP) and the Environmental Quality Incentives Program (EQIP). Congress authorized these programs in the 1985 and 1996 Farm Bills, respectively, as a result of increasing concern for conservation and water quality that had been building since the 1960s (Batie et al., 1985). The more recent Conservation Security Program (CSP) is a stewardship program that complements the CRP and EQIP. These programs are administered by USDA's Farm Service Agency (FSA) and its Natural Resources Conservation Service (NRCS), respectively.

The CRP provides technical and financial resources to assist eligible farmers and ranchers in improving soil and water management practices on their lands. The program initially focused on retirement of highly erosive and other environmentally sensitive land from crop production. However, the scope of the CRP has been steadily expanded, such that it now encompasses a broad range of natural resource management issues (SWCS, 2003). Total land area in the CRP is about 35 million acres. The CRP provides contracts under which producers receive rental payments for lands in the program. After the initial sign-ups under the 1985 Farm Bill, the USDA used this program to retire productive land, both as a supply control measure during the farming financial crisis of the late 1980s and to remove environmentally sensitive land from production. Since 1997, there has been more emphasis on retiring fragile lands that, when taken out of production, would yield improvements in water quality and wildlife habitat. The CRP is the largest USDA-sponsored conservation program, and it has yielded multiple and substantial environmental benefits (National Audubon Society, 1995). For example, Box 6-1 describes a conservation reserve enhancement program developed in Illinois that leverages and extends the federal CRP program. Smaller programs, including the Wildlife Habitat Incentive Program, the Wetlands Reserve Program, and the Grassland Reserve Program, augment the CRP.

EQIP, the second-largest program by expenditure (but first in terms of number of participants and acres under contract), provides financial and technical assistance to farmers and ranchers to implement practices and build infrastructure primarily to improve water quality and reduce erosion. It is the main USDA program for protection of environmental quality on working land. The program aims to provide producers with assistance that promotes production and environmental quality protection and improvement as compatible goals. Farmers carry out EQIP activities according to a plan of operations that identifies practices the farmer will implement in order to address site-specific natural resource concerns in addition to production objectives. Plans are subject to NRCS technical standards adapted for local conditions. These plans must be approved by the local conservation district. The program is implemented through local conservation districts, but the program does not effectively target working lands that produce the highest rates of nutrient and sediment pollutant loads. Furthermore, the program lacks the coordination that would help it achieve a far greater impact (SWCS, 2007). The EQIP program has potential to be employed more effectively and to realize greater reductions in nonpoint source water pollution.

Introduced in the 2002 Farm Bill, the Conservation Security Program is designed to assist farmers in implementing conservation practices on a whole-farm planning basis. It is a stewardship program designed to improve environmental quality and natural resource condition in agricultural land-

BOX 6-1
Illinois Conservation Reserve Enhancement Program

The Illinois River is a major tributary of the Mississippi River, and its drainage basin covers a large portion of Illinois including most of the prime agricultural land in the state. Illinois has developed a Conservation Reserve Enhancement Program (CREP) to restore and protect large stretches of floodplain corridors both on the mainstem of the Illinois River and along the major tributaries. It is helping landowners, who have only been able to produce crops in the area once or twice in the last decade, to retire these lands from agricultural production. As part of the agreement with the USDA for administration of the Conservation Reserve Program, the state provides an additional incentive to landowners to extend the 15-year federal CRP for an additional 15 or 35 years, or as a permanent easement. The purpose of this state program is to provide long-term environmental benefits by allowing certain environmentally sensitive lands in the Illinois River watershed to be restored, enhanced, or protected over a period of time. The state's CREP portion is driven by locally led conservation efforts that show landowner support. This program is a vehicle for a partnership between landowners, governmental entities, and nongovernmental organizations in addressing watershed quality problems.

Of the 116,410 acres of land enrolled in the federal CRP program, 38.3 percent (44,549 acres) are also participating in the expanded state option in the Illinois River basin; 7.6 percent of participating acres have conservation programs extended to 30 years; 5.3 percent will be extended to 50 years; and 87.1 percent of the conservation acreage will be maintained in perpetuity. All of these expanded programs are within the Illinois River basin. To participate in the enhanced CREP program the state must match 20 percent of the federal program. To date, Illinois has spent more than $49 million on this initiative.

scapes, while also providing a source of income to producers. As producers increase the use of water quality and erosion control best management practices (BMPs), payments are increased (Box 6-2). The CSP is the most comprehensive working lands program to date, but it has operated with only a modest budget (SWCS, 2007). Like EQIP, the CSP has potential to help reduce nonpoint source water pollution.

The USDA sponsors significant land and water conservation programs that could help address nonpoint source water pollution in the Mississippi River basin. Participation is voluntary, but there are financial incentives to implement BMPs, as defined by the agency and local conservation districts. Because not all landforms, cropping patterns, and farm fields yield similar levels of nutrient and sediment loadings, *effectiveness and efficiency are increased when conservation programs are directed at farms and watersheds with the highest pollutant loadings*. The USDA, NRCS, and FSA could improve these conservation programs by better targeting them to the greatest sources of land degradation and water pollution (SWCS, 2007; SWCS and Environmental Defense, 2007). Stronger interagency coordination also would improve these programs.

As an example of an existing interagency cooperative conservation program, the USDA and the EPA currently participate in the Conservation Effects Assessment Project (CEAP). Along with the USDA and EPA, other program participants are the Army Corps of Engineers, the U.S. Fish and

BOX 6-2
Best Management Practices for Land and Water Conservation in Agriculture

A best management practice has been defined as a practice or combination of practices that represents the most technologically effective and economically feasible means of preventing or reducing the pollutant load generated by nonpoint sources to a level that meets water quality goals (USEPA, 1980). Examples of agricultural management practices for water quality protection include the following (SWCS, 2007):

- Conservation tillage—leaving crop residue on the soil surface to reduce runoff and soil erosion
- Crop nutrient management—optimizing nutrient inputs to ensure that sufficient nutrients are available to meet crop needs while reducing nutrient export from farm fields
- Pest management—use of methods to control insects, weeds, and pests below economically harmful levels while protecting water, soil, and air quality
- Conservation buffers—vegetation of water conveyance channels and areas along streams and ponds to serve as a barrier for capture of nutrients, sediments, and other pollutants in runoff
- Irrigation water management—applying irrigation water input to meet crop water demands while minimizing contamination of ground- and surface water
- Grazing management—control of grazing and browsing activities on pasture and ranch lands to minimize water quality impacts (e.g., through fencing along streams)
- Animal feeding operations management—control of runoff and waste storage and treatment to minimize impacts on water quality
- Erosion and sediment control—use of methods to minimize erosion and capture eroded soil in runoff from lands affected by agricultural production

The effectiveness of BMPs in agricultural settings is a subject of ongoing study. By most reports, the movement toward conservation tillage (no-till and low-till) in the Mississippi River states has realized some successes. For example, the Iowa River (a Mississippi River tributary) showed improvement in total suspended solids concentrations following the 1985 Farm Bill that encouraged such practices.

Wildlife Service, the U.S. Geological Survey (USGS), and several nongovernmental organizations. The CEAP began in 2003 as an effort to quantify the environmental benefits of conservation practices used by private landowners participating in select USDA conservation programs. An independent review of the CEAP strongly endorsed its purpose of helping to implement existing conservation programs and design new ones, while offering recommendations for program improvement (SWCS, 2006). Although the CEAP may require some changes and adjustments to help achieve its program goals, the coordination it has promoted serves as an example of interagency initiatives that could improve water quality in the Mississippi River and the Gulf of Mexico (NRCS, 2007).

Building on the cooperative efforts within the CEAP, the USDA and EPA could extend their collaborative efforts to other areas of water quality management and monitoring. For example, the USDA and the EPA could strengthen their collaborative activities to help improve targeting of funds expended in the CRP, EQIP, and CSP programs. The EPA and the USDA could work together with conservation districts, extension agents, and farmers on programs such as water quality monitoring and alternative cropping practices. Ideally, this cooperation would result in better-targeted expenditures and programs that would help farmers improve economic profitability and also help realize water quality and related environmental improvements. At a larger scale, Mississippi River system-wide water quality monitoring is important to evaluating water quality impacts of the CSP, CRP, and EQIP programs.

KEY POLLUTANTS AND STRATEGIES FOR REDUCING THEIR IMPACTS

Nutrients

The nutrients of major concern with respect to the water quality of the Mississippi River and the Gulf of Mexico are nitrogen and phosphorus, especially from agricultural lands used for row crop production. In developing strategies for nutrient management in agricultural production, meeting essential nutritional needs for crops and livestock, producing profitable economic returns, sustaining environmental quality, and conserving natural resources are all important considerations. Effectively reducing nutrient impacts on Mississippi River basin water quality will require improved nutrient management strategies that balance nutrient requirements for crop production with reductions of nutrient loss from agricultural lands to surrounding watersheds.

Nitrogen

The challenge of meeting nutrient needs for crop production has resulted in an increasing demand for nutrients (fertilizer) to produce higher crop yields. When natural processes in the soil can no longer supply sufficient nutrients to meet crop production needs, farmers have applied increasing amounts of nutrients as fertilizers to agricultural lands. Meeting crop demands for nutrients such as nitrogen without causing loss of excess nitrogen to the environment is difficult because nitrogen undergoes continuous cyclic transformations into various forms and states in nature (Keeney and Hatfield, 2001). This "nitrogen cycle" results in many complicated spatial and temporal changes in the distribution of various nitrogen compounds in the environment. Nitrogen- and phosphorus-containing fertilizers may result in increased crop yields and economic return, but these additions also alter the distribution of various forms of nitrogen and phosphorus in the soil and can result in leaching and runoff of excess nitrogen and phosphorus to waterways (see Box 6-3).

Over the years, many Corn Belt states have used different approaches to develop nitrogen fertilizer application guidelines. This has inadvertently resulted in confusion among the Corn Belt states regarding appropriate fertilizer application rates. In recent years, many scientists from the upper midwestern states have noted that rates of nitrogen application needed to reach specific corn production yield goals are relatively consistent over this broad geographic region, but there are large variations in soil and climatic conditions and in management practices. This realization has led to the development of a regional approach for setting nitrogen application rate guidelines (Sawyer et al., 2006). The ability to set guidelines for optimal nitrogen application is important in the management of nitrogen-bearing fertilizer for water quality protection.

Although setting application rate guidelines is a critical step in developing a reliable management strategy for nitrogen, this approach only addresses the issue of how much nitrogen farmers should apply for optimal production. Soil testing alone cannot improve the efficiency of nitrogen use in crop production. Therefore, many states have also developed regional nutrient management guidelines. These guidelines include fertilization practices such as the timing and type of fertilizer nitrogen applications; tillage practices such as no-till, minimum tillage, conservation tillage; and use of cover crops (see Randall and Mulla, 2001; Randall and Vetsch, 2005). Furthermore, recent developments in precision agriculture technology have further enhanced the farmer's ability to manage more accurate and timely nitrogen applications (Mamo et al., 2003). These BMPs have been adopted widely for more efficient production of the predominant corn-soybean crop mix in the upper Midwest. These BMPs aim to increase agricultural

BOX 6-3
Fate of Applied Nitrogen and Phosphorus in Agricultural Soils

Under aerobic conditions, nitrate is normally the most dominant form of available nitrogen in the soil for crop production. Nitrogen in fertilizer, often in the form of anhydrous ammonia, is readily hydrolyzed into ammonium and subsequently oxidized into nitrate in the soil. During a growing season, the nitrogen available in the soil at any given time could be derived from fertilizers, manure, or composted organic wastes from various sources applied to the soil; from mineralization of soil organic matter, crop residues, or fertilizer residues from the previous cropping season; or by other means such as deposition from the atmosphere and biological nitrogen fixation. At the same time, microbial transformations, movement and leaching from soil, immobilization, denitrification, and nitrate reduction processes, in addition to crop uptake, reduce the amount of nitrate present in the soil. From the point of view of water quality and the potential for nitrogen pollution of the river, the form of nitrogen that is of major concern is also nitrate, because it is the form that is carried by water in runoff from soil surface or by leaching through the soil into the river or groundwater. However, from the viewpoint of the total quantity of nitrogen in the soil, nitrate is only a small component, with the vast majority of nitrogen present in organic forms. Nitrate is formed continuously from organic nitrogen, with the transformation affected by variations in soil physical properties, in temperature and available moisture during the growing season, and in other factors that influence nitrogen transformation processes in the soil.

Phosphorus is the other major essential nutrient needed for crop production that has caused significant concern because of its impact on the water quality of the Mississippi River (Wortman et al., 2005). The mechanisms and processes involved in its transport and transformation in the soil environment are different from those for nitrogen. The dominant form of phosphorus in the soil environment is phosphate. However, at any given time, only a tiny fraction of total soil phosphorus exists as phosphate ion in solution. The vast portion of soil phosphate exists as highly insoluble phosphate minerals (e.g., calcium, iron, and aluminum phosphates), tied strongly to soil clay particles or bound in soil organic matter. Unlike leaching of nitrate from soil, phosphorus is lost from land mostly through surface runoff carrying excess water and eroded soil particles and organic materials into the nearby river as suspended solids or sediments.

production, but the guidelines also protect environmental quality and can be incorporated into management practices to help meet Clean Water Act goals. The NRCS efforts to implement such BMPs could influence Mississippi River water quality in a positive way and should be combined with coordination and targeting of efforts under the CRP, EQIP, and CSP programs discussed earlier.

Phosphorus

Strategies for managing phosphorus, both for enhancing crop production and for preventing deterioration of water quality, are different from those for nitrogen. Most of the productive agricultural soils in the Midwest now contain high levels of phosphorus from years of application of manufactured fertilizers, manure, and biosolids (sludge). As a result, the potential for phosphorus pollution from surface water runoff is high, especially from fields devoted to row crops that have little plant residues covering the land surface. BMPs for these fields generally seek to limit external phosphorus inputs to the soil, maintain sufficient ground cover or crop residues on the soil surface to reduce soil erosion, and build buffer strips between crop fields and nearby rivers and streams to trap sediments and prevent them from entering surface and groundwater. Effective soil conservation practices are especially important in minimizing soil erosion on steeper fields.

Although phosphorus BMPs are, in principle, beneficial to both agricultural production and environmental quality, their effectiveness is difficult to evaluate at the farm field or local watershed level. Much of the phosphorus is particle associated. There is a considerable lag time between changes in soil management practices and improved water quality in rivers (Mulla et al., 2005). The limited amount of long-term water quality data to assess BMP effectiveness in improving environmental quality has confounded meaningful evaluation of the success of these BMPs in improving downstream water quality.

Nutrient management is a critical factor in agricultural production as well as in maintaining water quality, and farmers and government agencies must implement appropriate nutrient management strategies as part of a comprehensive and integrated approach to modern farming operations. Existing USDA conservation programs, especially EQIP and CSP, provide vehicles for doing just this and could be utilized more fully to help improve water quality across the Mississippi River basin and in the Gulf of Mexico.

Sediments

Agricultural activities result in enhanced sediment inputs to the Mississippi River, but the extent of agricultural contribution in a particular watershed is difficult to measure. Because of the nonpoint source nature of sediment pollutants, it is difficult to trace these pollutants back to their source. Even if a source location can be identified, it is challenging to assess quantitatively the extent of the pollution. For example, soil erosion can be an obvious source of sediments from a field, especially if the erosion process forms gullies and rills. However, sheet erosion is less visible but may carry more soil mass off the field. Differences in the extent of soil erosion occurring from field to field, and at various times during the year, can be significant because many different factors affect soil erosion. These include soil properties, fertility status and fertilizer applications, orientation and slope of the land, position of the field on the landscape (especially in relation to nearby streams), crops grown and cropping sequences, soil management practices, soil conservation measures, climatic conditions, use of irrigation, and crop growth stage during the growing season. It would be impractical to monitor continuously the amount of sediments coming off each farm field. Besides, the magnitude of soil erosion from a field may not be correlated directly with increase in sediments in nearby streams. Nevertheless, sediment inputs from agricultural lands can be estimated by a combination of measurement and modeling.

One approach to identify sources and reduce inputs of nutrients and sediments that contribute to water quality deterioration is used in the Minnesota River (Box 6-4). This effort involves a partnership between the State of Minnesota and a research institute at the University of Minnesota, which convenes and integrates a wide range of expertise to perform complex assessments and modeling of agricultural practices, soil and nutrient fluxes, and water quality impacts. The study team is evaluating nonpoint pollution sources and developing plans for nonpoint source reduction through a Total Maximum Daily Load (TMDL) framework. The project also illustrates the need for coordination between water quality regulatory authorities and agricultural agencies.

As in the Minnesota River watershed, variations in soil types, landforms, crop types, agricultural practices, and other factors result in regional differences in sediment and nitrogen fluxes. Achieving water quality standards and other water-related goals in the Mississippi River basin will require the identification and targeting of those subwatersheds that contribute most of the sediments and nutrients to the mainstem of the Mississippi River and its tributaries.

Targeting of USDA conservation programs can encourage farmers to implement BMPs for sediment and water runoff control on lands that are the primary sources of nonpoint pollutants. This process provides an opportunity to strengthen EPA-USDA interagency collaboration: the EPA can assist USDA in identifying lands with priority, and the EPA can cooperate with USDA and farmers in monitoring changes in water quality and making subsequent adjustments and improvements to nutrient management programs. The USGS could also play an important role in this collaboration by lending its considerable expertise and data related to water quality monitoring.

BOX 6-4
Evaluating Sediment Loadings to the Minnesota River

The Minnesota River recently has been studied intensively because of its contributions of sediments and phosphorus into Lake Pepin, a lake on the main-stem Mississippi River. Lessons from studies on the Minnesota River illustrate the complexity of the issues involved in assessing sediment loads and strategies needed to deal with sediment problems.

The Minnesota River watershed covers 10 million acres and contains 12 major watersheds. Monitoring "typical" branch watersheds as indicators for water quality impairments from fields within a watershed proved difficult because data collected from one watershed could not be extrapolated to others. Thus, agroecoregions were established in the Minnesota River basin by grouping farm land of similar landscape characteristics, cropping systems, and climatic regimes across tributary watersheds into various management units that farmers can readily identify (Hatch et al., 2001). This has resulted in the establishment of 13 agroecoregions in the Minnesota River watershed, a more manageable number for formulating BMP recommendations for farmers in each region (see *http://www.soils.umn.edu/research/mn-river/doc*). Since these initial efforts, agroecoregions have been delineated in other watersheds (see *http://www.soils.umn.edu/research/soilandwater_quality.php*).

Three of these watersheds—the Blue Earth, Le Sueur, and the Lower Minnesota River—cover 25 percent of the total area of the Minnesota River watershed, but contribute 66 percent of the sediment load. From 40 to 60 percent of the sediment in these watersheds has been estimated to arise from natural processes of stream bank or bluff erosion along the river channels (Sekely et al., 2002). The Minnesota Pollution Control Agency has been working with researchers at the University of Minnesota to develop TMDL guidelines and best management practices for agriculture in these three tributary watersheds. The efforts devoted to improve water quality along the Minnesota River reveal both the complexity of the problems in assessing sediment contributions from farm fields to nearby streams and the potential for developing appropriate methods to minimize runoff of sediments and nutrients from farm fields into nearby waterways.

APPROACHES FOR REDUCING NONPOINT SOURCE INPUTS FROM AGRICULTURAL LANDS

Targeting and Water Quality Improvement

Economists and other have argued for years that increased targeting—or focusing efforts in conservation, agricultural practices, and other practices on specific fields and farms—would improve effectiveness of conservation programs (Ribaudo, 1986; Wu and Boggess, 1999). The philosophy that underpins targeting is based on the fact that in some watersheds, a small fraction of land may cause the bulk of the nutrient and sediment loading.

Basing conservation and land use decisions across a watershed primarily on incentive-based payments to enlist voluntary actions does not ensure efficient use of resources designed to reduce nutrient and sediment loadings.

Specific watershed targeting of conservation programs for agriculture allows the relevant agencies to more efficiently deploy land-water conservation resources and expertise toward protecting and improving water quality in high-priority locations. For example, nutrient loadings to the Mississippi River are higher in the upper Mississippi River Corn Belt region than in lower portions of the river basin. The USDA could direct some EQIP and CSP funding to priority areas of the upper Mississippi River region to reduce nutrient loadings. The issue is the application of the appropriate program and the most appropriate action under that program for the pollutant of interest.

In terms of geographical targeting, riparian land and livestock grazing land have been the focus of special attention, and several different approaches to reduce pollution have been taken. Wetlands or riparian land (land adjoining water) can act as buffers against nutrients and sediments reaching water. Several USDA conservation programs encourage the creation of riparian buffers. There have been strong industry-government partnerships to promote riparian buffers and put them in place. However, landowners can find it challenging to preserve their continuing effectiveness. The challenge is to design incentives that encourage efficient management of the buffer over time.

Landowners can be paid under USDA programs to protect stream banks and limit livestock access to streams. In some locations, discharge from livestock management facilities and lands is a significant pollutant input to waterbodies (Kaufman and Kreuger, 1984). In addition, when livestock have access to rivers and other waterbodies, they can damage riparian zone vegetation and affect stream bank stability. Programs to reduce livestock access to streams have yielded significant water quality improvements and, if implemented at larger scales, can produce large-scale benefits. One example of extensive geographical targeting to reduce livestock impacts on water quality is the effort by New York City to reduce pollution in the upstream watershed region in the Catskills (Pires, 2004).

Political pressure tends to limit the extent to which conservation programs are targeted. Programs that target conservation assistance to particular geographic areas or enterprises are seen by some as unfair because not all producers can receive conservation payments. Reversing a trend that had been growing since the 1985 Farm Act, the 2002 Farm Bill excluded the opportunity to target on the basis of cost-effectiveness, but the administration's current proposal for the 2007 Farm Bill moves modestly toward allowing more targeting (USDA, 2007a). Although opposition to targeting is understandable, the fact remains that at the watershed or river basin level, some areas produce greater sediment and nutrient loadings than others. Distribution of the limited resources available for watershed-level nutrient and sediment management must use some criteria regarding effectiveness if agriculture-related programs are to offer an efficient means of improving water quality in the Mississippi River and the Gulf of Mexico.

Market-Based Approaches and Regulation

Targeting can be integrated into the different institutional approaches aimed at improving water quality, and the USDA has done some of this integration in the past. Market-based approaches, whether based on performance or design, can provide incentives to concentrate efforts. Performance-based approaches require monitoring and information that allows increased targeting. Both auction-based approaches and easements (which can be auction-based) are amenable to various degrees of targeting. Thus, limits on targeting derive primarily from lack of information or lack of political will.

Traditionally, regulators have relied on directives to mitigate pollution. All levels of government increasingly are tending to augment this approach, referred to as "command-and-control," through market-based policies. In market-based approaches to pollution control, a regulator sufficiently alters the relative value of available options for an individual polluter such that subsequent decisions have market incentives to align with the public or regulatory objective (Stavins, 2001). A well-designed market-based policy instrument often can accomplish the desired regulatory goal at comparatively lower cost than command-and-control regulation. In addition, market-based policies can provide significant incentives for cost-effective innovation that reduces abatement costs to the polluter and to society. The evolution of market-based strategies is a continuous process. A variety of market-based policy initiatives have been proposed in response to diverse situations, and there is no one standard approach. Although market-based incentives can be useful in promoting agriculture efficiencies and environmental improvements, they do not necessarily represent a panacea, and their successes depend on unique political, geographic, social, and economic contexts (see Devendra et al., 2006, for a summary of market-based approaches). The following section describes some commonly attempted market-based approaches.

Water Quality Trading

In conjunction with its watershed initiative, the EPA introduced a Water Quality Trading Policy in January 2003 (USEPA, 2003f). This market-based approach to improving water quality allows point sources and nonpoint sources—especially sources of nutrients (nitrogen and phosphorus) and sediment—to trade discharge allowances within areas of a watershed governed by an approved TMDL (USEPA, 2003f). Participants must possess a Clean Water Act permit, and the trade must result in improvements beyond those already achievable through the technology-based effluent limitations (USEPA, 2003f). Water quality trading is in its initial phases, but the program clearly contemplates cross-border trading and hence, logically, cross-border TMDLs. Of the 10 mainstem Mississippi River states, only Minnesota is currently experimenting with a trading program (USEPA, 2006f). Beyond the mainstem Mississippi River, other states have implemented different trading programs to help address water quality problems (see, for example, Box 6-5 for a discussion of nutrient trading in Pennsyslvania).

Water quality trading is a broad concept embracing a variety of compliance options for point and nonpoint sources under the Clean Water Act. In theory, a trading program allows parties to discharge pollutants up to some quota or limit. Those parties that discharge less than their allocated limit would generate credits that could be sold—and purchased by those parties that discharge pollutants beyond their allocated limit. Those who discharge beyond their limits have the choice of either reducing discharges or purchasing credits from the lower polluters. Theoretically, overall pollutants can be reduced, at lower social and economic costs, if (1) the aggregate limit of total pollution represents a reduction and (2) pollution control costs are met largely by those who have lower costs of pollution control. The realities of water quality trading, however, are more complicated. For example, existing National Pollutant Discharge Elimination System (NPDES) regulations do not allow dischargers to exceed permitted discharge. These types of regulatory and other realities pose significant complications to successful implementation of water quality trading programs.

Tradable permits have been used extensively for air pollution under 1990 amendments to the Clean Air Act. Air quality trading programs have seen some successes for a variety of reasons, one of which is that discharges are from point sources and can be measured and verified relatively easily, and the medium of trade is a standard "commodity" such as a ton of sulfur dioxide. Water quality trading in the Mississippi River basin would involve a large percentage of nonpoint dischargers, and air and water pollution issues fall under different statutory regimes—current statutory and regulatory constructs often make it difficult to structure effective, market-based trading programs (see Stephenson et al., 1999). Although the relative success of air quality trading permits should be considered, so should the significant differences between air and water quality trading regimes. There is an extensive literature on the realities, experiences, and pros and cons of implementing water quality trading (and TMDLs) that the interested practi-

BOX 6-5
Pennsylvania Nutrient Trading Program for the Susquehanna River

The Dauphin County Conservation District in Pennsylvania established a nutrient trading program available to Dauphin County farm owners. The program was created in response to a Pennsylvania Department of Environmental Protection (PADEP) initiative focused on enhancing the water quality of the Susquehanna River in order to meet federal mandates enacted to improve the health of the Chesapeake Bay. Pennsylvania has a comprehensive nutrient trading program related to water quality improvements in the Chesapeake Bay (see PADEP, 2007a).

Farmers accepted into the program receive cost-share funding to install selected agricultural best management practices, such as cover crops and no-till practices, to reduce the amount of nutrients in runoff from their lands. The installation of a BMP generates nutrient discharge trading credits that have monetary value. Different amounts of credit are linked to particular BMPs and their demonstrated effectiveness in reducing nutrient runoff.

Trading of the nutrient discharge credits allows point source dischargers, such as municipal wastewater treatment plants, to obtain nutrient reduction credits and thus meet their permit requirements. Credits are purchased from the agricultural nonpoint source dischargers and provide a source of income to the farmer. General guidelines for these transactions are that they must involve comparable units (e.g., nitrogen must be traded for nitrogen); they must be expressed as mass per unit time; they can occur only between eligible parties; credits generated by trading cannot be used to comply with existing technology-based effluent limits as expressly authorized by federal regulations; they may occur only in a watershed authorized by the PADEP; they are not allowed between sources outside of watershed boundaries; they may take place between any combinations of eligible point sources, nonpoint sources, and third parties; and each trading entity must meet applicable eligibility criteria established by the PADEP (2007b). In addition, all credits used to meet an annual nutrient cap, or any other effluent limitations, must be used under conditions contained in an NPDES permit. The Pennsylvania Department of Environmental Protection is responsible for program oversight and enforcement.

The two-year trial program is being implemented by the Dauphin County Conservation District, which is collaborating with PADEP (DCCD, 2007). It serves to illustrate not only a working nutrient trading program, but also what can be achieved through collaboration of state and federal water quality regulators with USDA and their conservation districts.

tioner or decision maker may wish to consult (see, for example, Stephenson and Shabman, 2001; Shabman et al., 2002).

Water quality trading programs face regulatory, monitoring, and other challenges. Nevertheless, water quality trading could become more useful and widespread over time as monitoring improves and as stricter water quality criteria are adopted (which has been the case for air pollution). Water quality trading may produce greater economic efficiencies, which could encourage additional future trading. These trading schemes also hold the prospect of providing multiple environmental benefits in the form of nonstructural, or "green," best management practices such as buffer strips, reforestation, constructed wetlands, and better fertilizer and other nutrient management practices. Meeting nutrient targets can be an expensive proposition, and water quality trading holds the prospect of a relatively low-cost means of helping meet these targets.

Performance-Based Trading

In some cases, nonpoint discharges can be measured accurately enough to allow actual performance to determine compliance with the cap-and-trade program rather than using estimates of performance from BMPs. For example, the Grass Lands Farmers' Trading Program in the San Joaquin Valley measures selenium discharges at the irrigation district level (Young and Karkoski, 2000). Trades are conducted among the seven irrigation districts. Each district has its own strategy to influence farmers within the district to reduce selenium loadings. Performance-based trading is usually easier with point sources, such as wastewater treatment plants or point source discharges from irrigation drainage tile systems, where monitoring and measurement of discharges are already required under the Clean Water Act's NPDES permit program.

Design-Based Trading

It is not always possible to determine accurately the extent of discharges from nonpoint sources such as agriculture. As a result, some watershed management authorities use a design-based water quality trading system instead. Under this framework, the nonpoint sources generate credits by adopting prescribed BMPs that are expected to reduce pollutants by a given amount. For example, the North Carolina Division of Water Quality, under its Tar-Pamlico Nutrient Reduction Trading Program, facilitated the formation of a consortium comprising both point and nonpoint sources to reduce nitrogen and phosphorous discharges (NCDENR, 1998). Point sources exceeding the limit can either invest in equipment to reduce their loadings or buy credits from farmers who have adopted nutrient-controlling BMPs (see Ribaudo et al., 1999, for further discussion of the characteristics of, and differences between, performance- and design-based approaches).

Auction-Based Contracting

Auction-based contracts determine which individuals are willing to undertake pollution control at what costs and can serve as a source of public information about pollution control. Citizens often lack information on the cost required to implement or maintain practices to reduce pollution. Traditional monetary incentive programs provide compensation to landholders for their efforts. Landowners may be overcompensated, however, if payments are substantially greater than the costs of the pollutant management measures (Stoneham et al., 2003). In addition, most conventional incentive programs do not recognize that different land segments differ with respect to their conservation significance or the synergies that can result from using multiple conservation strategies. Although some of these shortcomings are addressed in procedures such as the CRP auction process and the EQIP environmental benefit index, other kinds of auction-based contracting address them more successfully. For example, under the Australian Onkaparinga Catchment Water Management Board system, bidding is designed to limit as much as possible the landowner's knowledge of the board's willingness to pay (Brett et al., 2005). The closed-bid strategy with a limited number of contracts reveals the landowner's true costs; the selection of a bid based on the joint conservation significance of the land and the invested effort can result in a cost-effective allocation of public money. Such a strategy can allow precise targeting of resources to specific environmental concerns or multiple objectives (for more background on auction-based contracting, see Latacz-Loehmann and van der Hamsvoort, 1997).

Conservation Compliance

The 1985 Farm Bill introduced the concept of conservation compliance (Luzar, 1988). Under this management approach, for a producer to receive commodity price supports and other USDA program benefits, the producer would have to maintain certain conservation standards. These standards included both protection of existing wetlands and grasslands and the use of BMPs to keep soil erosion rates within set bounds. These standards have been relaxed since 1985. Enforcement was assigned to the Natural Resources Conservation Service and was extremely unpopular, effectively reducing its technical assistance role with producers (GAO, 2003; Wiebe and Gollehon, 2006). At issue is whether financial support for agricultural production also entails some responsibility for proper land stewardship. The Secretary of Agriculture's proposal for the 2007 Farm Bill would increase conservation compliance requirements (USDA, 2007a). High commodity prices, however, dull the effectiveness of conservation programs that are tied to price support payments.

MOTIVATING NONPOINT SOURCE CONTROL IN AGRICULTURE

A key factor in reducing nutrient and sediment pollution in the Mississippi River is the motivation of those who can control pollutant discharges. This degree of motivation is affected by a combination of institutional and economic considerations. Given the examples of market-based approaches, multiple incentives often are needed to produce outcomes that are both cost-effective and contribute to environmental protection or enhancement. Market-based approaches can become operative only if some enforceable regulatory standard provides the initial incentive to which market forces can respond. The institution providing the incentives also must have the appropriate geographical reach required to accomplish the pollution reduction goals and adequate enforcement authority.

The primary means in the United States to control point source discharges has been Clean Water Act NPDES permits, but for nonpoint agricultural sources, states and the federal government have mostly encouraged voluntary control measures through economic incentives. Incentives have often taken the form of direct payments from the rest of society, such as payments to farmers to set aside land under the CRP or payments under the EQIP or CSP to implement nutrient management plans. Tax incentives or disincentives can also be used. The fact that the Clean Water Act does not require command-and-control legislation for nonpoint sources highlights the importance and potential of the funded USDA conservation programs in helping improve water quality in the Mississippi River, its tributaries, and the Gulf of Mexico. These incentive programs gain even more importance if the USDA Conservation Compliance rules are increasingly less effective.

Although participation by farmers and ranchers in the USDA programs is voluntary, these programs have no shortage of applicants. Farmers compare the value of the incentive(s) offered to the cost of meeting the standards and requirements necessary to obtain the incentive(s) and decide whether to participate. These costs include not only direct costs such as management time and establishment of ground cover, but also forgone opportunity costs that might be involved in production activities such as growing crops or grazing additional livestock.

Nonmonetary concerns are also a part of farmers' crop production and nutrient management decisions. Some farmers may be predisposed to participate or not based on attitudes or levels of formal education, and some may perceive higher benefits and lower costs for participation than other farmers. In addition, if a farmer or society views the incentive program's objective favorably, participation is more likely. For the entity providing the incentives, therefore, the question is how to set the incentives at levels sufficient to generate adequate participation, without overpaying. This valuation issue explains why there is increasing interest in devices such as auction-based payments, which enlist farmers predisposed to participate at lower incentive cost than those who have to be compensated more to participate.

The USDA land and water conservation programs have benefited farmers and ranchers and resulted in some environmental improvements (SWCS and Environmental Defense, 2007); however, better targeting will be necessary to realize further substantial improvements in water quality as it is affected by agriculture. The suite of USDA programs aimed at farmers and ranchers clearly needs to be applied more effectively in order to realize additional reductions in nonpoint source pollution in the Mississippi River basin (GAO, 2003).

Improved coordination between the USDA, the EPA, and the states clearly can achieve more effective management of nonpoint water pollution sources from agricultural lands. There exist good examples of where cooperation on farming systems, nutrient management, tillage practices, and water quality monitoring has yielded improvements in water quality. Illustrative of these from within the upper Mississippi River basin are the programs and activities promoted by the Iowa Soybean Association, or ISA (Box 6-6). The ISA is not a federal program, but it demonstrates the many linkages among agriculture and water quality, at different spatial scales, and how collaborative efforts among farmers and water quality experts can produce additional benefits for both agriculture and water quality.

POTENTIAL IMPACTS OF BIOFUELS PRODUCTION

The potential for additional nonpoint source pollution from the expansion of bioenergy crop production illustrates the need for improved nonpoint source pollution control. Expanded biofuel production, especially ethanol, has the capacity to increase both sediment and nutrient loadings in the Mississippi River. The key drivers of such increases are as follows:

- Ethanol plant construction and increased production of ethanol have greatly increased the demand for corn.
- Increased prices for corn and other substitute crops create strong production incentives and dilute the attractiveness of voluntary conservation payments. High corn prices also potentially reduce the influence of cross-compliance if farmers do not have to join price support programs. Corn prices increased from about $2 per bushel in the fall of 2006 to more than $4 per bushel in early 2007 (USDA, 2007b). This price increase is unprecedented and is being driven primarily by anticipated increases in the use of corn in ethanol production in 2007 and 2008.
- There likely will be increased land across the Mississippi River

BOX 6-6
The Iowa Soybean Association: Programs for Reducing Nonpoint Source Impacts on Water Quality

The Iowa Soybean Association, established in 1964, develops policies and programs designed to help farmers expand profit opportunities and operational efficiencies while promoting environmentally sensitive production methods. ISA is governed by an elected board of 21 volunteer farmers and serves about 6,000 members in Iowa. ISA sponsors initiatives designed to help improve production and profitability, including market development for soy foods, soy biodiesel and bio-based products, and an on-farm network that helps evaluate in-field products and practices. ISA's agronomic and environmental programs address whole farming systems, including nutrient management and pest control in corn and soybean production, integration of livestock and manure management in crop production, tillage practices, and energy management.

ISA environmental programs encompass three primary initiatives: Certified Environmental Management Systems for Agriculture (CEMSA), watershed management programming, and an On-Farm Network™. These initiatives aim to develop, apply, and promote programs that assist producers in increasing productivity and efficiency and that enhance agriculture's ability to measure and improve environmental performance. All rely on the principles and practices of *applied evaluation* (collection of site-specific data) and *adaptive management* (integration of data into management decisions for continual improvement).

The ISA watershed program involves planning at the watershed level and extends to include farm operational level issues and field-level considerations. ISA promotes a philosophy of integrating various activities among at least a majority of production acres across a given watershed in order to realize water quality gains. The goal of this philosophy is to improve sustainable production on working lands and further mitigate nonpoint source pollution through targeted placement of buffers and wetlands. ISA works with farmers to help gather and evaluate water quality data to characterize waters, identify trends over time, identify emerging problems, assess the effectiveness of control programs, and direct pollution control activities to areas in which they will have the greatest effect.

The On-Farm Network involves field trials of different management approaches for improved agricultural production and environmental performance. It provides a mechanism for testing and demonstration of best management practices. The program's main focus has been on nitrogen management in corn production. In the growing seasons since 2000 when the program began, ISA has coordinated field trials with participating farms to help reduce nitrogen application rates and modify nitrogen application timing, method, and form. Data from the field trials have been compiled and evaluated by ISA, with the results disseminated to farmers and state and federal agencies. The On-Farm Network program serves as an example of the kind of nonregulatory initiative for agricultural process improvement that can lead to reduced nonpoint source impacts on water quality.

SOURCE: Adapted, with permission, from Iowa Soybean Association (2007).

basin under cultivation, including potential CRP land going back into crop production to increase total crop acres. This possibility already concerns a number of wildlife groups interested in the wildlife benefits of CRP (Brasher, 2006). Moreover, the additional land that farmers would bring into production would be more marginal than lands currently in production. Spring 2007 planting intentions indicate more than 10 million additional acres of corn for 2007. This increase in corn will come primarily from decreased soybean acres, but also from decreases in acres planted in wheat and cotton. Continuous corn will replace corn-soybean rotations in many cases. While there was traditionally a 50-50 corn-soybean rotation in the upper Midwest, a 60-40 rotation is being projected. Greater nitrogen leaching from the increased corn production will be a major concern. This trend toward increased corn production is not limited to the Corn Belt region: large areas of agricultural land in the Mississippi River delta region are being converted from cotton to corn, for example, and acreage planted to corn is also projected to increase in some Eastern states.

- Increased continuous corn production, as opposed to traditional corn-soybean crop rotation, will have negative effects on water quality. To maintain yields that were achieved under traditional crop rotation practices, continuous corn production requires more fertilizer and often more erosive tillage systems (Vyn, 2007).

A large block of CRP contracts was due to expire in 2007 releasing land for possible crop production. Because of administrative staffing limitations, USDA decided to let farmers re-enroll land (ahead of contract expiration) that had contracts expiring in 2007-2010 for varying time periods if the land provided sufficiently high environmental benefits. Well over 80 percent of the 27.8 million acres with contracts expiring during this period were re-enrolled starting early in 2006. Much of the re-enrollment occurred before the tremendous run-up in corn prices during the 2006 fall harvest and subsequent high prices in 2007 that would have discouraged re-enrollment. Thus, only a small number of acres will be released from contract that might enter crop production from the CRP. There are currently some 4 million to 7 million acres that could support corn or soybean production now in the CRP that might come out eventually for that purpose.

SUMMARY

Runoff from agricultural lands is the primary nonpoint source of nutrients and sediments to the Mississippi River and the Gulf of Mexico. There is an inherent conflict between agricultural production and improving water quality in the Mississippi River. The USDA's traditional agricultural commodity programs tend to encourage more production and more intensive production.

Although the Clean Water Act does not authorize command-and-control regulation for nonpoint sources such as agricultural lands, the USDA has instituted programs to reduce the water quality impacts of agriculture. Through these programs, the USDA is the key organization in managing agricultural nonpoint source pollution. These voluntary, incentive-based programs include the Conservation Reserve Program, the Environmental Quality Incentive Program, and the Conservation Security Program. The programs aim to balance incentives for crop production with incentives for land and water conservation on farms and ranches. Participation is voluntary, but there are financial incentives for implementation of best management practices. The national financial investment in and the scope of these USDA programs is large. **It is imperative that these USDA conservation programs be aggressively targeted to help achieve water quality improvements in the Mississippi River and its tributaries.**

Current application of USDA environmental protection programs is not well targeted to the most significant sources of land degradation and water pollution, but targeting could be much improved through interagency coordination. Because not all farm fields across the Mississippi River basin contribute equal amounts of nutrients and sediments that eventually make their way to the river, water quality protection programs need not be implemented in every watershed and on every farm. **Programs aimed at reducing nutrient and sediment inputs should include efforts at targeting areas of higher nutrient and sediment deliveries to surface water.**

The EPA and the USDA should strengthen their cooperative activities designed to reduce impacts from agriculture on the water quality of the Mississippi River and the northern Gulf of Mexico. Management of nutrient and sediment water inputs and other water quality impacts will require site-specific, targeted approaches involving BMPs. Existing USDA programs provide vehicles for implementing agricultural nonpoint source controls, but they will require closer coordination with the EPA and state water quality agencies to maximize water quality improvements. The EPA could provide assistance to the USDA to help improve targeting of the significant funds expended in the CRP, EQIP, and CSP programs. The EPA and the USDA should draw on the considerable expertise and data of the USGS in implementing programs that include water quality monitoring components.

The prospects of greatly expanded bioenergy production and robust commodity markets are encouraging producers to extend and intensify crop production across the upper Mississippi River basin. Much of this expanded production is in corn, which entails high rates of fertilizer application and intensive soil tillage. As a result, nutrient and sediment runoff from agricultural land in the upper Mississippi River basin is likely to increase. This state of affairs provides an even stronger rationale to implement with urgency the targeted application of USDA conservation programs, to improve and expand EPA-USDA coordination for nonpoint pollution control programs, and to devise and implement other initiatives to mitigate the adverse effects of nutrients and sediments on the Mississippi River and the Gulf of Mexico.

7

Collaboration for Water Quality Improvement Along the Mississippi River Corridor

Management of water quality in interstate rivers under the Clean Water Act's framework poses challenges for both state and federal agencies tasked with implementation of the act. States have the primary responsibility for implementing most of the act's provisions through direct legislative authority or delegation of programs from the Environmental Protection Agency (EPA). These responsibilities include permitting, water quality standard development, monitoring, and where necessary, preparation of Total Maximum Daily Loads (TMDLs). Coordination and cooperation among states with shared surface waters is critical for effective water quality management under the Clean Water Act (USEPA, 1998b). The EPA also plays a major role through its mandated oversight to ensure that state programs for shared surface waters are compatible and consistent with goals of the Clean Water Act and related federal statutes. This role is particularly critical on the Mississippi River for which large-scale issues such as Gulf of Mexico hypoxia are linked with inputs and processes in upstream regions many hundreds of miles away.

Clean Water Act (CWA) implementation along the Mississippi River represents a substantial scientific and public administration challenge, because it requires some degree of coordination among the 31 basin states, especially the 10 mainstem states. It also requires coordination among several federal and state agencies and activities. The Mississippi River flows through four EPA regions, while seven EPA regions oversee water quality protection activities across the entire river basin. Because delivery of CWA water quality programs is ultimately the EPA's responsibility, coordination among the multiple EPA regions with Mississippi River basin jurisdiction, particularly in the 10 mainstem states, is crucial. Coordination among other federal agencies is also necessary, because the U.S. Army Corps of Engineers (USACE) and the U.S. Geological Survey (USGS) have CWA-related programs and responsibilities along the river, and the National Oceanic and Atmospheric Administration (NOAA) has some water quality monitoring responsibilities in the Gulf of Mexico.

This chapter examines programs under the Clean Water Act in which federal, interstate, and state-federal coordination is needed for effective water quality protection in the Mississippi River. It examines existing and potential collaborations among states, EPA regions, and other federal agencies pertaining to Clean Water Act implementation, and the experience of various organizations that have been established to facilitate state and federal coordination on other shared U.S. waters. Finally, the chapter assesses the potential for using some of the approaches adopted by these organizations as models for improving Mississippi River water quality management.

CLEAN WATER ACT COORDINATION NEEDS ON AN INTERSTATE RIVER

As this report has explained, the pillars of the Clean Water Act are effective National Pollutant Discharge Elimination System (NPDES) point source permitting programs that achieve best available treatment technology; water quality standards comprising designated uses and water quality criteria; adequate monitoring to ensure protection of water quality and achievement of water quality standards; assessment to evaluate water quality status; and restoration programs to improve waters with impaired water quality relative to designated uses. For interstate rivers, coordination among states is important for effective implementation of each of these Clean Water Act components.

Water quality standards are central to Clean Water Act implementation. As explained earlier in this report, states develop standards for particular waterbodies that consist of use designations and criteria for the waterbody's physical, chemical, and biological quality. States and the EPA use these standards to establish water quality-based effluent limitations for point source discharges, to assess surface water quality, and to develop restoration programs, based on TMDLs, for waterbodies that do not meet standards. Different use designations and associated water quality criteria established by different states for the same shared waterbody can lead to conflicts in permitting, monitoring programs, assessment conclusions, and restoration strategies.

Monitoring programs that both state and federal agencies administer are critical to the ability to determine the extent to which surface waters are meeting relevant water quality criteria, to understand trends and existing or emerging problem areas, and to assess progress toward achievement of water quality goals (Chapter 5). Disparate monitoring goals and methods promote fragmented data sets, inconsistent laboratory results, and an overall inability to define water quality problems accurately.

Assessments of water quality status determine the type of protective or restorative action needed for a particular waterbody. Unilateral assessments by individual states of shared waters can yield different conclusions regarding water quality for the same body of water. Such disparities are confusing to the public and promote inconsistent and even conflicting response programs for water quality remediation.

Under the Clean Water Act, for waterbodies that remain impaired after required point source controls are adopted, additional restoration plans must be developed. The primary corrective approach specified in the Clean Water Act is Section 303(d), which requires assessment, identification of impaired waterbodies, and development of TMDLs to address water quality impairments. Water quality improvements will be difficult to achieve if multiple states with jurisdiction over sources discharging to the same waterbody are not in agreement about allocation of pollutant loads. For the Mississippi River, multistate coordination is essential for any prospect of effective water quality protection and restoration.

COOPERATION ON INTERSTATE RIVERS

Although states historically have focused most of their time and resources on programs to protect waters within their own jurisdictions, some activities specified in the Clean Water Act have resulted in coordination among states for shared waters such as the Mississippi River. When states are drafting permits for point source discharges to waters with boundaries shared with or upstream of another state, the permitting state must forward those permits to the adjoining or downstream state for comment before the permit is finalized. If necessary, a public hearing process is available to resolve any disputes. Like permitting, standards development provides a mechanism for state-to-state interaction. Public notice and hearings are required for any revision to a state's water quality standards, including both water quality criteria and designated uses. These hearings provide an opportunity for adjoining states to raise concerns about interstate issues. Similar public notice provisions apply to the development and adoption of TMDLs for waters shared by or impacting those of another state. Although these CWA provisions offer an enforceable mechanism for states to interact on major decisions of joint interest, experience has shown that the mainstem Mississippi River states seldom use them.

Monitoring programs on shared waters constitute another area in which direct state-to-state cooperation is desirable but has been limited on the Mississippi River. Missouri and Iowa, for example, use monitoring data that Illinois collects in order to avoid duplication at prime monitoring sites (e.g., bridges), which enables them to redirect resources to other priorities. However, as noted in Chapter 5, these limited cooperative activities have done little to expand data collection efforts on the Mississippi River. Although direct state-to-state interaction is desirable in implementing the Clean Water Act—and some formal mechanisms exist to facilitate such exchanges—Mississippi River states have seldom used these mechanisms effectively because of the river's size and its numerous jurisdictions. A more structured approach is needed, such as a formal mechanism for cooperation among states for water management.

In other river systems, states have coordinated their activities for water management principally through interstate compacts. Moreover, Section 103 of the Clean Water Act requires the EPA administrator to "encourage cooperative activities by the States and encourage compacts between States for the prevention and reduction of pollution." This section further provides that "the consent of the Congress is hereby given to two or more States to negotiate and enter into agreements or compacts. . . ." Numerous compacts exist to help define the many aspects of managing water, such as allocations, standards, and responsibilities (Table 7-1). Some interstate and federal-interstate river basin commissions oversee and help implement the provisions of these interstate compacts. Of the river basin commissions established under compacts listed in Table 7-1, six of them receive funding under Section 106 of the Clean Water Act:

1. Delaware River Basin Commission
2. Interstate Commission on the Potomac River Basin
3. Interstate Environmental Commission (Tri-State Compact)
4. New England Interstate Water Pollution Control Commission
5. Ohio River Valley Water Sanitation Commission
6. Susquehanna River Basin Commission

Notably, all of the "Section 106" commissions were established prior to enactment of the 1972 Clean Water Act (UMRBA, 2006). Congress did not provide funding for similar commissions in the future. Four large river systems for which states have established compacts that include water quality management as an objective are the Delaware River, the Ohio River, the Potomac River, and the Susquehanna River.

Delaware River Basin Commission

The Delaware River Basin Commission (DRBC) was established in 1961 as part of the Delaware River Basin Compact, a federal-state com-

TABLE 7-1 Partial Listing of Interstate Compacts with Water-Related Provisions and Functions

Compact	Signatories	Objective
Saco Watershed Compact	New Hampshire, Maine	Watershed development
Bear River Compact	Idaho, Utah, Wyoming	Water allocation
California-Nevada Interstate Compact	California, Nevada	Equitable apportionment of water conservation, development
Colorado River Compact	Arizona, California, Colorado, Nevada, New Mexico, Utah, Wyoming	Water apportionment and development
Columbia River Compact	Washington. Oregon	Regulating, protecting, and preserving fish
Columbia River Gorge Compact	Washington, Oregon	Watershed development
Connecticut River Atlantic Salmon Compact	New Hampshire, Massachusetts, Connecticut, Vermont, U.S. Fish and Wildlife Service, National Marine Fisheries Service	Restoration of anadromous Atlantic salmon to the Connecticut River
Connecticut River Flood Control Compact	Vermont, Massachusetts, Connecticut, New Hampshire	Flood protection
Delaware River Basin Compact	New York, New Jersey, Delaware, Pennsylvania, United States	Protect, enhance, and develop water resources of the basin
Great Lakes Basin Compact	Indiana, Michigan, Minnesota, New York, Ohio, Pennsylvania, Wisconsin, Illinois	Development, conservation-balanced uses
Interstate Compact for Jurisdiction on the Colorado River	Arizona	Concurrent law enforcement
Interstate Compact on the Potomac River Basin	Maryland, Pennsylvania, Virginia, West Virginia, District of Columbia, United States	Water resources management and interstate pollution abatement
Interstate Public Water Supply Compact	New Hampshire, Vermont	Joint public water supply facilities
Kansas-Missouri Flood Protection and Control Compact	Kansas, Missouri	Prevention and control of floods
Klamath River Basin Compact	California, Oregon	Development, use, conservation
Merrimack River Flood Control Compact	Massachusetts, New Hampshire	Water storage, utilization, and flood control
Missouri River Barge Compact	Iowa, Kansas, Missouri, Nebraska	River development for barge traffic

TABLE 7-1 Continued

Compact	Signatories	Objective
New England Interstate Water Pollution Control Compact	Connecticut, Maine, Massachusetts, New Hampshire, New York, Rhode Island, Vermont	Abatement of interstate water pollution
Ohio River Valley Water Sanitation Compact	Illinois, Indiana, Kentucky, Ohio, West Virginia, Pennsylvania, New York, Virginia	Interstate water pollution control
Oregon-California Goose Lake Interstate Compact	California, Oregon	Basin development, water use, and conservation
Republican River Compact	Colorado, Kansas, Nebraska	Water allocation
Snake River Compact	Idaho, Wyoming	Development, use, flood protection
South Platte River Compact	Colorado, Nebraska	Water apportionment
Susquehanna River Basin Compact	New York, Pennsylvania, Maryland, United States	Water resources management
Tahoe Regional Planning Compact	California, Nevada	Conservation, preservation
Tri-State Compact	New York, New Jersey, Connecticut	Water and air pollution abatement
Upper Niobrara River Compact	Wyoming, Nebraska	Water apportionment, groundwater information
Yellowstone River Compact	Montana, North Dakota, Wyoming	Water apportionment, development

SOURCE: Reprinted, with permission from ICWP (2006). © 2006 by Interstate Council on Water Policy.

pact among of Delaware, New Jersey, Pennsylvania, New York, and the United States. The DRBC consists of the four basin state governors and a uniformed Corps of Engineers officer appointed by the President. The compact's objectives include facilitating interstate comity; providing for planning, management and control of water resources; providing for cooperative planning and action by the signatory parties; and applying the principle of equitable allocation. The commission's annual budget is approximately $4.5 million and it employs 42 full-time staff.

Although the DRBC works under the authority of its compact, several DRBC programs support Clean Water Act provisions. These programs include designating special protection waters, development of TMDLs, water quality and groundwater monitoring, biomonitoring, fish tissue analysis, ambient toxics and sediment surveys, and coordination of states' activities.

Further, the DRBC sets water quality standards and regulates effluents and water withdrawals. Among the commission's institutional and operational challenges is a funding shortfall from the federal government. Although the United States is a compact signatory and, as such, is obligated to help fund commission operations, Congress has not appropriated its contractually based share (20 percent) since 1997.

Ohio River Valley Water Sanitation Commission

The Ohio River Valley Water Sanitation Commission (ORSANCO) was established in the Ohio River Valley Water Sanitation Compact of 1948 to help abate interstate water pollution. Participants include the states along the Ohio River—Illinois, Indiana, Kentucky, Pennsylvania, Ohio, and West Virginia—and New York and Virginia, which lie within the watershed's upper reaches. Although the United States is not a signatory, there are three federal representatives on the commission along with three representatives from each state.

ORSANCO is empowered to establish treatment standards for waste discharges to interstate streams within the participating states' Ohio Valley drainage area, conduct surveys, recommend state legislation to achieve pollution abatement goals, and confer with any party that has an interest in water pollution control. ORSANCO's activities concerning Clean Water Act implementation include the adoption of water quality standards, permitting coordination, water quality and biological monitoring and assessment, TMDLs, and Gulf of Mexico hypoxia abatement. ORSANCO also coordinates CWA programs with the source water protection provisions of the Safe Drinking Water Act. ORSANCO's annual budget is approximately $3.5 million and it has a full time staff of 25 (UMRBA, 2006).

Interstate Commission on the Potomac River Basin

The Interstate Commission on the Potomac River Basin (ICPRB) was established in 1940 to assist Potomac River basin states and the federal government to enhance, protect, and conserve the basin's waters and associated land resources. The ICPRB comprises three commissioners and three alternate commissioners from Maryland, West Virginia, Pennsylvania, Virginia, and the District of Columbia, along with three presidential appointees. ICPRB's annual budget is approximately $2.4 million, which supports a staff of 23 (UMRBA, 2006). Programs related to Clean Water Act implementation include the Chesapeake Bay Program (see Chapter 4), TMDLs, spill modeling and tracking, water quality monitoring, and assessment and evaluation of indicators. Like the DRBC, Congress currently is failing in its funding obligation to the ICPRB.

Susquehanna River Basin Commission

The Susquehanna River Basin Commission (SRBC) was established in 1970 for management of water resources in the Susquehanna River basin. Maryland, New York, Pennsylvania, and the United States are compact members. The commission has one representative (the governor) from each participating state and one federal representative who is appointed by the President (the current federal representative is from the U.S. Army Corps of Engineers).

The SRBC focuses on flood mitigation and management of water resources for municipal, agricultural, recreational, commercial, and industrial purposes, but water quality protection and restoration are also part of its mission. The commission has undertaken various water quality monitoring, assessment, and restoration programs and participates in the EPA-coordinated, multistate effort to protect and restore water quality in the Chesapeake Bay. The SRBC's annual budget is approximately $4.5 million and it has a full-time staff of 34. Congress currently is not providing its full funding obligation to the SRBC.

Prospects for Mississippi River Compacts

Clean Water Act implementation for the Mississippi River has not yielded the type and extent of state cooperation and coordination that have been achieved for the Delaware, Ohio, Potomac, and Susquehanna Rivers. In the case of DRBC, ORSANCO, ICPRB, and SRBC, each agency's programs reflect the coordination needs unique to each river. DRBC programs that delineate special protection waters may provide a model with regard to the Mississippi River for areas of special ecological significance. ORSANCO's organization of numerous committees of state and federal agency clean water program management personnel may also serve as a model for some Mississippi River administrative issues. These committees meet under ORSANCO's aegis to discuss coordination needs and also to design and implement programs that eliminate duplication of effort, including development of databases and assessments, plans for response to spills, and integration of Safe Drinking Water Act (SDWA) and Clean Water Act requirements. ICPRB and SRBC have been engaged by their cooperating states in coordinating roles to assist with reduction of nutrients to support restoration of the Chesapeake Bay, thus illustrating how an interstate organization can affect the type of cooperation needed to address nutrient pollution problems in a river system's estuarine and gulf areas.

An interstate compact can be an effective approach to water quality management, but can take many years to establish. Historically, the average time to enact the 19 existing compacts that govern river management and water rights has been approximately nine years (UMRBA, 2006). Experience has shown that congressional consent for an interstate compact, although not a legal requirement, is desirable to help protect the compact from invalidation by a future act of Congress (UMRBA, 2006). Compacts are difficult to establish today because of complexities in creating an agreeable compact, resistance to the ceding of state authority to an interstate entity, and difficulty of obtaining long-term state and federal funding commitments.

A 2006 report from the Interstate Council on Water Policy listed several principles that would help support programs for interstate water quality management programs (Box 7-1). These principles may be relevant for the Mississippi River states if they are considering possible future organizational and administrative frameworks to help improve interstate water quality management.

COOPERATIVE EFFORTS ALONG THE MISSISSIPPI RIVER

A compact is not the only mechanism for enhancing cooperation among states in management of shared waters, and the mainstem Mississippi River states have undertaken several non-compact, cooperative efforts focused on Mississippi River water quality management since passage of the Clean Water Act. Some of these have limited goals, whereas others are aimed at broader goals and long-term planning and cooperation. Most of the initiatives have achieved at least partial success but have encountered substantial obstacles to efficient, effective, and sustained collaboration and cooperation. Such obstacles include resource constraints, competing priorities among participating states, confusion regarding regulatory primacy for shared waters, and technical challenges related to monitoring water quality in a large, interstate river. This section reviews the experiences with interstate cooperation for water quality management along the Mississippi River.

Early Efforts

Initiatives to improve interstate cooperation along the Mississippi River have taken place for many years in the contexts of navigation and flood control (Anfinson, 2003; ICWP, 2006; UMRBA, 2006). Recognizing the need for improved and more systematic state-state and state-federal cooperation to improve water quantity and quality management, Congress passed the Water Resources Planning Act in 1965. Title II of this legislation authorized a series of federal-state river basin commissions and provided financial assistance to states for comprehensive river basin planning. In response, several river basin commissions were established across the United

BOX 7-1
Principles of Interstate Cooperation for Water Quality Management

The effectiveness of interstate compacts or agreements was recently evaluated by the Interstate Council on Water Policy (ICWP, 2006). In that study, the following characteristics were identified that provide a "compelling rationale of such institutional arrangements":

- **Strength in numbers and enhanced voice:** Multijurisdictional arrangements, such as interstate commissions (among many other forms), provide individual members with an opportunity to speak and act with a single, harmonized voice.
- **Monitoring and surveillance:** Ecosystem assessment programs provide the science-based data and information critical to program design, implementation, and evaluation. Such programs can be prohibitively expensive for a single jurisdiction, and to maximize their value, they need to be implemented on a watershed basis.
- **Pooling and accessing resources and expertise:** Multijurisdictional institutional arrangements allow individual members to leverage limited resources to dramatically increase capability in areas such as assessment, research, program design and implementation, and policy development, among others.
- **Ecosystem-based management:** Now widely accepted as a fundamental operating principle, the ecosystem approach to resource management recognizes the interrelatedness of ecosystem components and an associated need for a comprehensive, integrated, and multimedia management strategy.
- **Regional priority setting:** Individual institutions operating within a watershed find that inefficiency and unwanted redundancy can be avoided through a single priority setting process.
- **Communication, collaboration, and technology transfer:** Information exchange with like-minded professionals enhances efficiency, fosters partnerships, and encourages the type of innovation and creative thinking needed to advance the practice of watershed-based resource management.
- **Uniformity, consistency, and program effectiveness:** Results can be negated or otherwise compromised due to inconsistencies in the way multiple jurisdictions within a single watershed address a shared issue (e.g., pollution sources, fishing limits, invasive species prevention and control).
- **Protecting jurisdictional interests:** Jurisdictions can participate in a multijurisdictional institution as a means of "keeping an eye on" neighboring jurisdictions and other parties that may have goals contrary to their own. In the course of evaluating the pros and cons of various alternatives, each participant gets a chance to understand and demonstrate respect for the needs and contributions of other communities.

States, including the Upper Mississippi River Basin Commission. However, in 1981, President Reagan issued Executive Order 12319 calling for the dismantling of the Water Resources Planning Act commissions. This order ended federal support for these Title II commissions, but states preserved many of them in some form in order to maintain their interstate planning and coordination services. For example, the Upper Mississippi River Basin Association (UMRBA) was formed in response to the termination of the Upper Mississippi River Basin Commission.

A planning and coordination commission was not developed for the lower Mississippi River under the Water Resources Planning Act of 1965. In that part of the basin, efforts remained focused on navigation and flood control, with the federal government represented by the U.S. Army Corps of Engineers in the lead role. Water quality issues were secondary at best and, in many respects, remain of low priority today because of the limited ability of the lower Mississippi River states to influence water quality in the river (Chapter 5). However, there were some efforts to initiate interstate water quality coordination in the lower Mississippi River in the period after the federally supported commissions ended.

In 1987, for example, the Louisiana legislature passed legislation directing the governor to execute a Lower Mississippi River Pollution Phase-out Compact with the United States and the upstream states along the river's course. This was an initiative of the Louisiana Department of Environmental Quality. The compact's intended purposes were to reduce and then eliminate river pollution by 1998; encourage alternatives to discharging wastes and pollutants into the river; and maintain the biological and chemical integrity of the Mississippi River system to ensure water quality adequate for drinking water, agricultural, aquaculture, and recreational uses. In addition, the compact sought to ensure the collection and sharing of information among the signatories relative to technologies, methods, incentives, and regulatory means that could improve Mississippi River water quality. This compact was never put in place, however.

Cooperation Outside of Compacts

Upper Mississippi River Basin Association

As noted above, the Upper Mississippi River Basin Association was established in 1981 as a successor to the Upper Mississippi River Basin Commission. A joint resolution signed by the governors of Illinois, Iowa, Minnesota, Missouri, and Wisconsin created the UMRBA and called for "the continuation of cooperation of the interstate organization to maintain communication and cooperation among the states on matters related to water planning and management." Gubernatorial appointees, generally individuals from state agencies with substantial responsibilities for water resource management, represent UMRBA member states. Certain federal agencies participate as advisers to the UMRBA, including the U.S. Army Corps of Engineers, the U.S. Department of Agriculture (USDA) Natural Resources Conservation Service (NRCS), the Coast Guard, the Fish and Wildlife Service, the U.S. Geological Survey, the Maritime Administration, and the EPA.

Although the UMRBA encourages dialogue and coordination of Clean Water Act program activities for the upper Mississippi River states, its role is purely advisory and its resources are modest. The upper Mississippi River states have recognized the need to create a stronger collaborative and cooperative mechanism for water quality management and, accordingly, are working through UMRBA to assess the feasibility of establishing an interstate organizational structure with the capacity to coordinate and/or administer water quality programs under the Clean Water Act (UMRBA, 2006). The six interstate commissions that receive federal funding under Section 106 of the Clean Water Act were examined in detail as part of this assessment. The UMRBA has developed a plan for phased expansion of its role toward becoming a body to help administer interstate water quality programs (UMRBA, 2006).

An interstate compact is one approach to expanding both the authority of and the resources available to the UMRBA for improved implementation of the Clean Water Act on the Mississippi. However, such an approach will require significant dedicated funding from participating states, as well as formal legislative action by each state to ratify a compact. Given the limited state resources and the time needed to pursue the necessary legislation, a formal compact is not planned for the near term but remains a possibility for the future (UMRBA, 2006).

Upper Mississippi River Water Suppliers Coalition

The Upper Mississippi River Water Suppliers Coalition (UMRWSC) was established in 1999 to serve as a focal point to represent the common interests of the drinking and industrial water suppliers and to establish a formal communication network for the membership (UMRWSC, 2006). The organization also serves as a resource clearinghouse for river water quality and related information, promotes source water protection practices pursuant to the SDWA, and provides educational opportunities for members and their customers. The UMRWSC is working toward developing and maintaining an early-warning source water monitoring network by developing working relationships with other river stakeholders, particularly on river water quality initiatives. The UMRWSC has 26 water supplier members and holds periodic meetings. UMRWSC water suppliers collect water quality data for the upper Mississippi River on a daily basis, and their combined long-term data record is an important resource.

Upper Mississippi River Sub-basin Hypoxia Nutrient Committee

Following up on a recommendation in the 2001 Hypoxia Action Plan (USEPA, 2001) to address hypoxic conditions created by Mississippi River discharge into the Gulf of Mexico, the states in the upper Mississippi River subbasin formed a committee to examine the relationship of agricultural practices to Gulf hypoxia. The Upper Mississippi River Sub-basin Hypoxia Nutrient Committee (UMRSHNC) includes the Illinois Department of Agriculture, the Iowa Department of Agriculture and Land Stewardship, the Minnesota Pollution Control Agency, the Missouri Department of Natural Resources, and the Wisconsin Department of Natural Resources. The subbasin committee, through its state members, is positioned to identify key stakeholders who need to be involved in the development and implementation of strategies to reduce nutrient loads to the Gulf of Mexico and to waterbodies within the basin. In 2004, the UMRSHNC formed a stakeholder group that includes representatives of key agricultural and environmental organizations and municipal, state, and federal agencies. UMRSHNC intends its activities to achieve a near-term goal of a technically sound and economically viable nutrient reduction strategy for the upper Mississippi River subbasin, and a long-term goal of reducing nutrient loadings to streams and lakes within the five states and to the northern Gulf of Mexico.

Upper Mississippi River Conservation Committee

The Upper Mississippi River Conservation Committee (UMRCC) is a cooperative, nonprofit organization of state and federal agencies formed to address the challenges of protecting and restoring the natural resources of the upper Mississippi River. Founded in 1943, the organization involves representatives of Iowa, Illinois, Minnesota, Missouri, and Wisconsin, along with the U.S. Fish and Wildlife Service (USFWS), USACE, and various local pollution control agencies. The UMRCC aims to provide continuing cooperation between conservation agencies responsible for fish, wildlife, and recreational management on the upper Mississippi River.

Lower Mississippi River Conservation Committee

The Lower Mississippi River Conservation Committee (LMRCC) is a cooperative, nonprofit organization of state and federal agencies formed to address the challenges of protecting and restoring the natural resources of the lower Mississippi River. The LMRCC, founded in 1994, is focused on coordination of state and federal efforts. It has no regulatory authority, but it provides a regular forum for discussion of water quality and natural resource protection and restoration issues. Participants include representatives of the environmental quality and natural resource agencies of each of the six lower Mississippi River states—Arkansas, Kentucky, Louisiana, Mississippi, Missouri, Tennessee—and representatives of five federal agencies—EPA, NRCS, USACE, USGS, and the USFWS.

The mission of LMRCC encompasses the full spectrum of natural resources linked to the river. Thus, the organization does not focus exclusively on water quality. In 2000, the LMRCC completed a Lower Mississippi River Aquatic Resource Management Plan, a 10-year operational plan to address the primary factors adversely affecting aquatic resources in the river's active floodplain and backwater areas. The LMRCC recognizes the need for a comprehensive system-wide assessment, such as that recommended under Section 102 of the Clean Water Act, and it is working to organize a lower Mississippi River resource assessment (Nassar, 2006). The LMRCC also is working to encourage the EPA to include the lower Mississippi River in the Environmental Monitoring and Assessment Program (EMAP), similar to the upper Mississippi River EMAP effort (Ingram, 2006).

Lower Mississippi River Sub-basin Committee on Gulf Hypoxia

A Lower Mississippi River Sub-basin Committee on Gulf of Mexico hypoxia, similar to that for the upper Mississippi River subbasin, was formed in 2003 to support the Hypoxia Action Plan (USEPA, 2001). This committee expects to coordinate the Hypoxia Action Plan's implementation in the lower river basin and to work with other states to ensure federal funding. Participating states are Arkansas, Louisiana, Mississippi, Missouri, and Tennessee. Regarding implementation, the committee will compile information on nutrient loadings (Mississippi and Atchafalaya River basins); assess impacts of state and federal programs aimed at reducing loadings; coordinate interstate watershed programs; promote and coordinate complementary regional and state efforts; and establish an open process for interested stakeholders, partner agencies, and universities to participate in and support pollutant reduction programs.

Nongovernmental Organizations

There are many nongovernmental organizations that focus on water resources, watershed protection, and water quality in the Mississippi River and its tributaries. These organizations typically focus on the waters of particular states. Examples include the Louisiana Environmental Action

Network, the Tennessee Clean Water Network, the Iowa Environmental Council, the Minnesota Center for Environmental Advocacy. Some of the organizations promote interstate cooperation and data sharing, such as the Mississippi River Basin Alliance, a coalition of more than 80 environmental organizations and conservation groups with interests related to the Mississippi River. At the scale of the entire river, the Mississippi River Water Quality Collaborative, which is sponsored by the McKnight Foundation, brings together representatives from more than 20 nongovernmental organizations from states along the Mississippi River corridor to explore strategies for comprehensive, riverwide water quality improvements.

EPA COLLABORATION ON THE MISSISSIPPI RIVER

The EPA has a major role to play in the implementation of the Clean Water Act in the Mississippi River basin both directly through the development of guidance documents such as water quality criteria and oversight of programs that it has delegated to the states and indirectly through regional coordination of programs to protect major rivers. The Mississippi River watershed encompasses seven EPA regions Regions 2, 3, 4, 5, 6, 7, and 8. Historically, each EPA region has been delegated and has exercised considerable discretion in carrying out the various federal programs entrusted to the EPA, including the NPDES and Section 303(d) (TMDL) programs. As Figure 7-1 shows, four EPA regional offices share responsibility for the mainstem of the Mississippi River. These EPA offices have their headquarters in Atlanta (Region 4), Chicago (Region 5), Dallas (Region 6), and Kansas City (Region 7).

Inconsistency among regional offices has been a persistent problem for the EPA with regard to a variety of Clean Water Act issues, including determination of impaired waters, approval of revised water quality standards, and enforcement. Such inconsistencies arise from several factors, including a lack of clear guidance from EPA headquarters, inexperienced personnel, and differing views of appropriate federal-state relations (GAO, 2000b).

The St. Louis Compact of 1997 involved six EPA regions, the EPA Office of Water, and the EPA Gulf of Mexico Program, and sought to increase coordination of programs and activities in the Mississippi River Basin. The goals of the compact were to improve coordination and communication among regions, to develop the capacity to evaluate natural resource economic issues and address hypoxia issues in the Gulf of Mexico, and to characterize the basin's physical and ecological features. Mississippi River basin strategy teams were to carry out these activities. Region 5 was designated the "lead" for the upper Mississippi River and Region 6 for the lower Mississippi. Although strategy team efforts ensued over several years following the signing of the agreement, activities under its aegis appear to have waned.

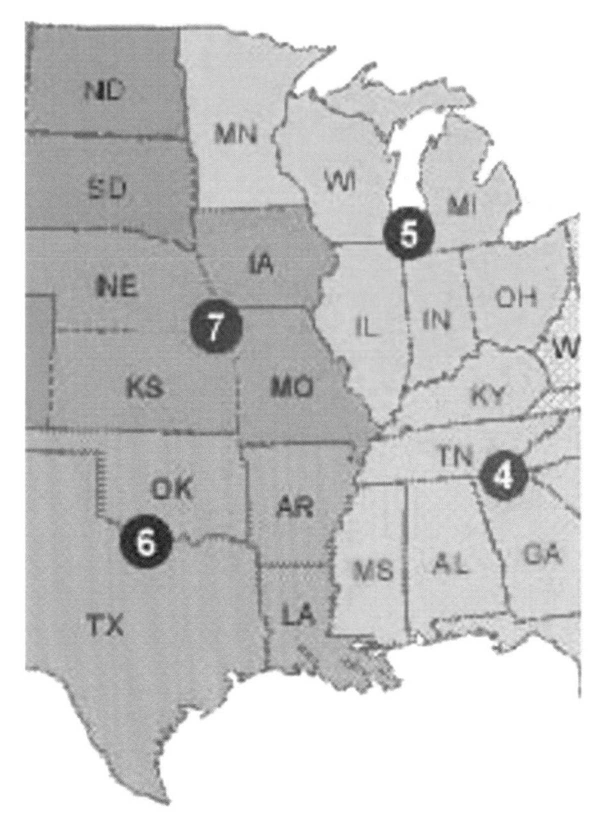

FIGURE 7-1 EPA regions with responsibility for the mainstem Mississippi River. SOURCE: Adapted from USEPA (2007e).

The EPA regions must work with the states within their respective jurisdictions regarding implementation of Clean Water Act programs for which states have direct responsibility and those delegated to them by the EPA, both on the tributaries to the Mississippi River and, in principle, on the Mississippi River itself. The EPA has the authority, although it rarely exercises that authority, to review and approve discharge permits prior to their issuance. EPA must review state water quality standards periodically for their adequacy and approve any changes to existing standards that the states propose. In addition, the EPA requires states to submit an annual (or semiannual) program plan describing state water quality activities in support of the Clean Water Act as a condition of federal funding. The EPA provides a major source of funding to state agencies to implement the Clean Water Act. As such, it can influence significantly state priorities. The EPA also interacts with watershed associations, river basin associations, river commissions, and other regional and interstate groups that work on water quality issues for the Mississippi River and its tributaries. These activities include the provision of funds to help support several of the interstate organizations discussed above.

There has been limited intra-agency coordination among the relevant EPA regions regarding Mississippi River water quality issues; instead, the various EPA regions focus primarily on interactions with states on water quality issues and waterbodies of primary concern to the states. Some coordination is taking place, however. For example, in the upper Mississippi River basin, EPA Regions 5 and 7 are working with the UMRBA to improve coordination on water quality management issues (UMRBA, 2004). In 2001, the UMRBA undertook an upper Mississippi River water quality coordination project designed to identify and explain the approaches that each upper Mississippi River basin state uses in its Clean Water Act Section 305(b) assessments and Section 303(d) impaired waters designations. The results from this UMRBA effort illustrate the collaborative contributions of EPA Regions 5 and 7 (UMRBA, 2004). On the other hand, in the lower Mississippi River basin, EPA Regions 4 and 6, which encompass the states bordering the lower Mississippi River, appear to have had only limited involvement with the LMRCC.

The EPA has recognized the importance of expanded and improved intra-agency coordination for more effective management of water quality in the Mississippi River. This large river usually receives only secondary attention from the states bordering it (as discussed in Chapter 4).

Stronger leadership by the EPA in promoting interstate and interregion cooperation on Mississippi River water quality monitoring issues—including development of specific water quality criteria documents tailored to the unique needs of the Mississippi River—could promote a common framework for coordinating key water quality protection programs for the river and the northern Gulf of Mexico. The EPA has the legal authority to create a watershed-wide entity to ensure adequate protection of Mississippi River water quality. The EPA could, for example, create a coordinating program office for water quality in the Mississippi River, comprised of representatives of the four EPA regions that encompass the Mississippi River mainstem, plus the EPA Gulf of Mexico Program office, to ensure that these criteria are integrated into state programs (as discussed in Chapter 4, a relevant model of interregional cooperation and EPA coordination is the Chesapeake Bay Program). Whatever approach is taken, the EPA clearly must assume a strong coordinating role to ensure water quality protection and improvement in the Mississippi River and the northern Gulf of Mexico. Commensurate with a stronger coordinating role would be stronger efforts by EPA to promote better cooperation among the 10 Mississippi River mainstem states, which is consistent with and complementary to a "water-

BOX 7-2
EPA and the Watershed Approach

The Environmental Protection Agency encourages citizens, agencies, and nongovernmental organizations to view water management as something that is most appropriately conducted at the watershed scale. EPA has promoted the watershed approach as "the most effective framework to address today's water resource challenges" since the early 1990s. EPA considers this approach to be hydrologically defined, inclusive of all stakeholders, and able to strategically pursue water resources goals. The approach is one of the four pillars of EPA's own Sustainable Infrastructure Initiative. EPA supports several web sites devoted to watershed planning and management and has issued several documents that explain its vision for watershed-level management and offer planning tools for watershed management.

Given the river's role in supporting interstate commerce and the river's ecosystems that cross state lines and extend into the Gulf of Mexico, federal interest in the Mississippi River is undeniable. Managing Mississippi River water quality among the multiple states along its corridor, and across its river basin area, is watershed planning on the largest scale and would be consistent with EPA's promotion of watershed-scale programs and initiatives.

SOURCE: USEPA (2007f).

shed approach" to water management. EPA has been vigorously promoting this watershed approach for more than 10 years (see Box 7-2), and stronger efforts by EPA in promoting interstate collaboration along the Mississippi River fit well with this paradigm. Furthermore, several National Research Council (NRC) reports issued in the 1990s and 2000s encourage the use of the watershed and the river basin as management units to help promote better management of drinking water supplies, environmental goods and services, and drought and water conservation programs (NRC, 1999a, 1999b, 2000b, 2002, 2004a, 2004b, 2007).

COOPERATION AMONG FEDERAL AGENCIES ON THE MISSISSIPPI RIVER

Numerous federal agencies have jurisdiction over activities that influence water quality in the Mississippi River. Of primary importance in this regard are the Environmental Protection Agency, with its responsibilities under the Clean Water Act; the U.S. Army Corps of Engineers, with its authorizations related to water resource management for navigation and flood control; and the U.S. Department of Agriculture, with its activities under the Farm Bill to minimize impacts of agricultural practices on water quality.

The U.S. Fish and Wildlife Service is an important actor through its authority and responsibilities regarding the Endangered Species Act. The U.S. Geological Survey also plays a significant role through its streamflow and water quality monitoring activities throughout the Mississippi River basin. Finally, the National Oceanic and Atmospheric Administration is a crucial participant given its monitoring responsibilities in the Gulf of Mexico.

With respect to water quality protection in the mainstem Mississippi River, a prominent example of interagency coordination is the Mississippi River-Gulf of Mexico Watershed Nutrient Task Force. The task force includes states, tribes, and federal agencies and was established in 1997 to evaluate and consider options to mitigate hypoxic conditions in the northern Gulf of Mexico, where the Mississippi River discharges its nutrient load. The EPA has a leadership role in this effort. Table 7-2 lists state and federal agencies that participate in the task force and the two subbasin committees that have been established: the Upper Mississippi River Sub-basin Hypoxia Nutrient Committee and the Lower Mississippi River Sub-basin Committee on Gulf Hypoxia. Federal agencies involved are those with responsibilities

TABLE 7-2 Federal and State Participants in the Mississippi River-Gulf of Mexico Watershed Nutrient Task Force

	Subbasin Committees	
Federal Agencies	Upper Mississippi River Sub-basin Hypoxia Nutrient Committee	Lower Mississippi River Sub-basin Committee on Gulf Hypoxia
Council on Environmental Quality	Illinois Department of Agriculture	Arkansas Soil and Water Conservation Commission
National Oceanic and Atmospheric Administration	Iowa Department of Agriculture and Land Stewardship, Soil Conservation Division	Louisiana Department of Environmental Quality
U.S. Army Corps of Engineers	Minnesota Pollution Control Agency	Louisiana Governor's Office of Coastal Activities
U.S. Department of Agriculture	Missouri Department of Natural Resources	Mississippi Department of Environmental Quality
Natural Resources Conservation Service	Wisconsin Natural Resources Department	Missouri Department of Natural Resources
U.S. Department of Justice		Tennessee Department of Agriculture
U.S. Department of the Interior		
U.S. Environmental Protection Agency		
U.S. Fish and Wildlife Service		
U.S. Geological Survey		
White House Office of Science and Technology Policy		

for activities in the Mississippi River basin, the Louisiana coastline, and the Gulf of Mexico. States in the broader Mississippi River basin also participate, although not all basin states are represented. For the Ohio River valley, ORSANCO serves as the coordinating agency for all activities of the Ohio Basin Subcommittee. Thus, ORSANCO, its member states, and Tennessee cooperate on Gulf of Mexico hypoxia-related issues.

In 2001 the Mississippi River-Gulf of Mexico Watershed Nutrient Task Force issued an action plan (USEPA, 2001) for reducing, mitigating, and controlling hypoxia in the northern Gulf of Mexico. The plan outlined a range of possible actions to reduce nutrient loads, with a focus on nitrogen, and to increase nitrogen retention and denitrification in the Mississippi River and its tributaries. Progress on the activities recommended in the plan has been limited. As of 2007, the action plan was under a five-year assessment that was one year past due. In a June 2007 letter to the EPA, a group of 11 Gulf of Mexico scientists noted this limited progress on the action plan and on addressing the hypoxia problem: "It is now nearly halfway between the submission of the *Action Plan* to the President and the Congress in January 2001 and the 2015 target date for reducing the hypoxic zone to less than 5,000 km^2. Yet there is no evidence of progress toward that goal and modest implementation of actions to achieve it" (UMCES, 2007).

With a continued interest in a watershed-based, "ecosystem partnership" approach for water quality management that emerged in the 1980s (ICWP, 2006), both opportunities and the need for cooperation among federal agencies for water quality protection and restoration in large river systems are greater than ever. With the authority created by the Clean Water Act and the continuing mission expansion of other federal agencies such as the Corps of Engineers and the USDA to encompass water quality, the EPA is well positioned to lead cooperative efforts among federal agencies. Such efforts can produce innovative approaches, better leveraging of available resources, and significant water quality improvements.

The length of the Mississippi River, the numerous states along its corridor, the river's importance in interstate commerce, and the ecosystems that span several states all justify a strong federal role for coordinated, rational, and effective management of water quality. On the basis of its authority under the Clean Water Act, its regional organization, and its relationships with state water quality agencies, the EPA clearly is the federal agency in the best position to provide this needed coordination and management guidance. The EPA should exercise a stronger coordinating role in improving interstate cooperation and consistency in water quality standards, monitoring, and control. Several ongoing activities could be expanded to good effect (e.g., the National Stream Quality Accounting Network program discussed in Chapter 5). There are also opportunities for new cooperative efforts among the states that EPA is well positioned to lead or assist. In making progress on Mississippi River water quality management issues, it will be important to take stock of existing programs and ensure that future efforts effectively draw from, and do not duplicate, existing efforts.

SUMMARY

Implementing the Clean Water Act for water quality protection and improvement along the 10-state Mississippi River corridor is a complex and challenging endeavor. At present, it is not being carried out effectively because of inadequate coordination among state and federal agencies. Successful implementation of the Clean Water Act for the Mississippi River requires improved coordination on every level. There is cooperation among various groups regarding Mississippi River mainstem water quality issues. However, with the exception of the Upper Mississippi River Basin Association, these collaborations generally do not focus on CWA implementation issues. For example, states may share some information gained in their monitoring activities, but they are not collaborating on design and implementation of Clean Water Act program components, such as the development of water quality standards and TMDLs.

There is a strong need for improved regional cooperation and coordination on water quality issues for the lower Mississippi River states, where progress generally lags behind that seen in the upper river states. Although many aspects of the UMRBA experience, including its organizational structure, may not directly and immediately transfer to the lower basin states, these states would benefit from more formal and stronger cooperative efforts. One option for promoting better lower river cooperation would be to provide additional resources and responsibilities to the existing Lower Mississippi River Conservation Commission. Another option would be to establish an interstate water quality body as part of an interstate compact, as has been done with organizations such as the Delaware River Basin Commission and ORSANCO. A third option would be to establish a non-compact organization such as the UMRBA which offers an advantage in that it could be established relatively quickly. Further, UMRBA represents an existing model, on the same river, that has been beneficial in many ways and could be replicated for the states of the lower river. **Better interstate cooperation on lower Mississippi River water quality issues is necessary to achieve water quality improvements. The lower Mississippi River states should strive to create a cooperative mechanism, similar in organization to the UMRBA, in order to promote better interstate collaboration on lower Mississippi River water quality issues.**

The upper and lower portions of the Mississippi River have very different features and therefore present distinctive water quality issues and challenges. At the same time, improved management of basinwide water quality issues, such as excess nutrient loading, requires more active coordination, and there is a distinct need for integrated management of water quality by all 10 states. For example, periodic meetings involving upper and lower Mississippi River water quality professionals, which could be convened by the EPA, would strengthen communication and collaboration among the 10 river states. Integration of water quality-related activities along the entire Mississippi River will require better coordination among the 10 Mississippi River mainstem states. The states will achieve far more by working cooperatively than by each state's going it alone.

The EPA should encourage and support the efforts of all 10 Mississippi River states to effect regional coordination on water quality monitoring and planning and should facilitate stronger integration of state-level programs. The EPA has an opportunity to broker better interstate collaboration and thereby improve delivery of Clean Water Act-related programs, such as permitting, monitoring and assessment, and water quality standards development. The EPA should provide a commensurate level of resources to help realize this better coordination.

Better consistency and integration of Mississippi River water quality programs is inhibited by the fact that four EPA regions have responsibilities for different stretches of the Mississippi River. Cooperation regarding water quality standards and TMDL development is an example of intra-agency coordination that would yield immediate benefits for Mississippi River water quality management. Indeed, the EPA has recognized the importance of intra-agency coordination for more effective management of water quality in the Mississippi—but it has so far failed to meet this challenge effectively. A useful model of regional cooperation and EPA coordination, as explained in Chapter 4 of this report, is the Chesapeake Bay Program. Whatever the approach adopted, a strong EPA coordinating role is essential for water quality protection and improvement in the Mississippi River. **The EPA administrator should ensure coordination among the four EPA regions along the Mississippi River corridor so that the regional offices act consistently with regard to water quality issues along the Mississippi River and in the northern Gulf of Mexico.**

References

Alexander, R. B., and R. A. Smith. 2006. Trends in the nutrient enrichment of U.S. rivers during the late 20th century and their relation to changes in probable stream trophic conditions. Limnology and Oceanography 51: 639-654.

Alexander, R. B., A. S. Ludtke, K. K. Fitzgerald, and T. L. Schertz. 1997. Data from Selected U.S. Geological Survey National Stream Water-Quality Monitoring Networks (WQN) on CD-ROM. Open-File Report 96-337. Available online at http://pubs.usgs.gov/dds/wqn96cd/html/report/contents.htm.

Alexander, R. B., R. A. Smith, and G. E. Schwarz. 2000. Effect of stream channel size on the delivery of nitrogen to the Gulf of Mexico. Nature 403: 758-761.

Anfinson, J. O. 2003. The River We Have Wrought: A History of the Upper Mississippi. Minneapolis, Minn.: University of Minnesota Press.

Antweiler, R., D. Goolsby, and H. Taylor. 1995. Nutrients in the Mississippi River. In R. Meade, (ed.), Contaminants in the Mississippi River. U.S. Geological Survey Circular 1133. Denver, Colo.

Arkansas Department of Environmental Quality (ADEQ). 2006. State Permits Branch. Available online at http://www.adeq.state.ar.us/water/branch_permits/default.htm#401.

Aulenbach, B. T., H. T. Buxton, W. A. Battaglin, and R. H. Coupe. 2007. Streamflow and Nutrient Fluxes of the Mississippi-Atchafalaya River Basin and Subbasins for the Period of Record Through 2005. U.S. Geological Survey Open-File Report 2007-1080. Summary available online at http://toxics.usgs.gov/highlights/of-2007-1080.html.

Barber, L. B., II, J. A. Leenheer, W. E. Pereira, T. L. Noyes, G. A. Brown, C. F. Tabor, and J. H. Writer. 1995. Organic contamination of the Mississippi River from municipal and industrial wastewater. In Contaminants in the Mississippi River. R. Meade (ed.). U.S. Geological Survey Circular 1133. Denver, Colo.

Barbour, M. T., J. Gerritsen, B. D. Snyder, and J. B. Stribling. 1999. Rapid Bioassessment Protocols for Use in Streams and Wadeable Rivers: Periphyton, Benthic Macroinvertebrates, and Fish, 2nd edition. U.S. EPA 841 B-99-002. Washington, D.C.: U.S. Environmental Protection Agency, Office of Water.

Barras, J. A. 2006. Land Area Change in Coastal Louisiana after the 2005 Hurricanes—A Series of Three Maps. U.S. Geological Survey Open-File Report 2006-1274. Available online at http://pubs.usgs.gov/of/2006/1274/.

Barry, J. M. 1997. Rising Tide: The Great Mississippi Flood of 1927 and How It Changed America. New York: Simon and Schuster.

Batie, S., L. Shabman, and R. Kramer. 1985. U.S. agricultural and natural resource policy: Past and future. In The Dilemmas of Choice. K. Price (ed.). Washington, D.C.: Resources for the Future.

Bennett, E. M., S. R. Carpenter, and N. F. Caraco. 2001. Human impact on erodable phosphorus and eutrophication: A global perspective. BioScience 51: 227-234.

Boesch, D. F. 2002. Challenges and opportunities for science in reducing nutrient over-enrichment of coastal ecosystems. Estuaries 25: 744-758.

Boesch, D. F., M. N. Josselyn, A. J. Mehta, J. T. Morris, W. K. Nuttle, C. A. Simenstad, and D. J. P. Swift. 1994. Scientific assessment of coastal wetland loss, restoration and management in Louisiana. Journal of Coastal Research, Special Issue 20: 1-103.

Boyd, G. R., and D. A. Grimm. 2001. Occurrence of pharmaceutical contaminants and screening of treatment alternatives for Southeastern Louisiana. Annals of the New York Academy of Sciences 948(1): 80-89.

Boyer, H. A. 1984. Trace elements in the water, sediments, and fish of the Upper Mississippi River, Twin Cities metropolitan area. Pp. 195-230 in Contaminants in the Upper Mississippi River. J. G. Wiener, R. V. Anderson, and D. R. McConville (eds.). Boston: Butterworth Publishers.

Brasher, P. 2006. Use of CRP Land for Ethanol Scares Wildlife Groups. Des Moines, Iowa: Des Moines Register.

Bratkovich, A., S. P. Dinnel, and D. A. Goolsby. 1994. Variability and prediction of freshwater and nitrate fluxes for the Louisiana-Texas shelf: Mississippi and Atchafalaya River source functions. Estuaries 17: 766-778.

Brett, B., S. Gatti, J. Connor, M. Garrod, and D. King. 2005. Catchment Care—Developing an Auction Process for Biodiversity and Water Quality Gains. A NAP Market-Based Instrument Pilot Project. Australia: Policy and Economic Research Unit, CSIRO Land and Water and Onkaparinga Catchment Water Management Board.

Burkholder, J. M., M. A. Mallin, and H. B. Glasgow, Jr. 1999. Fish kills, bottom-water hypoxia, and the toxic *Pfiesteria* complex in the Neuse River and estuary. Marine Ecology Progress Series 179: 301-310.

Caffey, R. H., P. Coreil, and D. Demcheck (eds.). 2002. Mississippi River Water Quality: Implications for Coastal Restoration. Interpretive Topic Series on Coastal Wetland Restoration in Louisiana, Coastal Wetland Planning, Protection, and Restoration Act, National Sea Grant Library No. LSU-G-02-002, 4 pp.

Cain, Z., and S. Lovejoy. 2004. History and Outlook for Farm Bill Conservation Programs. Choices, 4th Quarter. Available online at http://www.choicesmagazine.org/2004-4/policy/2004-4-09.htm.

Chavas, J., and M. Holt. 1990. Acreage decisions under risk: The case of corn and soybeans. American Journal of Agricultural Economics 72(3): 529-538.

Chesapeake Bay Memorandum of Understanding. 2000. Memorandum of Understanding Among the State of Delaware, the District of Columbia, the State of Maryland, the State of New York, the Commonwealth of Pennsylvania, the Commonwealth of Virginia, the State of West Virginia, and the United States Environmental Protection Agency Regarding Cooperative Efforts for the Protection of the Chesapeake Bay and Its Rivers. Available online at http://www.chesapeakebay.net/pubs/sixstatemou.pdf.

Chesapeake Bay Program. 2003. Setting and Allocating New Cap Loads. Available online at http://www.chesapeakebay.net/info/wqcriteriatech/newcap.cfm.

Chesney, E. J., and D. M. Baltz. 2001. The effects of hypoxia on the northern gulf of Mexico coastal ecosystem: A fisheries perspective. Pp. 321-354 in Coastal Hypoxia: Consequences for Living Resources and Ecosystems. N. N. Rabalais and R. Eugene Turner (eds.). Coastal and Estuarine Studies 58. Washington, D.C.: American Geophysical Union.

City of Minneapolis. Undated. Combined Sewer Overflow—A Minneapolis Solution. Available online at http://www.ci.minneapolis.mn.us/cso/ (includes CSO annual reports for 2003-2006).

Clean Water Act. 2006. 33 U.S.C. Section 1251-1387.

Clean Water Restoration Act (CWRA) of 1966. Pub. L. No. 89-753, 80 Stat. 1246, 1250.

Committee on Environment and Natural Resources (CENR). 2000. Integrated Assessment of Hypoxia in the Northern Gulf of Mexico. Washington, D.C.: National Science and Technology Council.

Cooper, C. M., and J. R. McHenry. 1989. Sediment accumulation and its effects on a Mississippi River oxbow lake. Environmental Geology and Water Sciences 13: 33-37.

Corbett, K. T. 1997. Draining the metropolis: The politics of sewers in nineteenth century St. Louis. Pp. 107-125 in Common Fields: An Environmental History of St. Louis. Missouri Historical Society Press.

Craig, J. K., and L. B. Crowder. 2005. Hypoxia-induced habitat shifts and energetic consequences in Atlantic croaker and brown shrimp on the Gulf of Mexico shelf. Marine Ecology Progress Series 204: 79-94.

Craig, J. K., T. A. Henwood, and L. B. Crowder. 2005. Effects of abundance and large-scale hypoxia on the spatial distribution of brown shrimp (*Farfantepenaeus aztecus*) on the northwestern Gulf of Mexico continental shelf. Canadian Journal of Fisheries and Aquatic Sciences 62: 1295-1308.

Craig, R. K. 2004. The Clean Water Act and the Constitution: Legal Structure and the Public's Right to a Clean and Healthy Environment. Washington, D.C.: Environmental Law Institute.

Dahl, T. E. 1990. Wetlands Losses in the United States 1780s to 1980s. Washington, D.C.: U.S. Department of the Interior, Fish and Wildlife Service.

Dauphin County Conservation District (DCCD). 2007. Agricultural Programs and Services. Available online at http://www.dauphincd.org/main/agservices.php.

Delaney, R. L., and M. R. Craig. 1997. Longitudinal Changes in Mississippi River Floodplain Structure. U.S. Geological Survey, Project Status Report, PSR 97-02.

Department of the Interior (DOI). 2007. Upper Mississippi River National Wildlife and Fish Refuge, U.S. Department of the Interior, Fish and Wildlife Service. Available online at http://www.fws.gov/Midwest/UpperMississippiRiver/.

Devendra, C., P. Bala, and O. Doering. 2006. Market Based Policy Instruments in Natural Resource Conservation. Report to the Natural Resources Conservation Service, Resource Economics and Social Sciences Division, Washington, D.C.

Diaz, R. J., and R. Rosenberg. 1995. Marine benthic hypoxia: A review of its ecological effects and the behavioural responses of benthic macrofauna. Oceanography and Marine Biology: An Annual Review 33: 245-303.

Donner, S. D., M. T. Coe, J. D. Lenters, T. E. Twine, and J. A. Foley. 2002. Modeling the impact of hydrological changes on nitrate transport in the Mississippi River Basin from 1955 to 1994. Global Biochemical Cycles 16:10.1029/2001GB001396.

Downing, J. A., J. L. Baker, R. J. Diaz, T. Prato, N. N. Rabalais, and R. J. Zimmerman. 1999. Gulf of Mexico Hypoxia: Land-Sea Interactions. Council for Agricultural Science and Technology, Task Force Report No. 134. 40 pp.

Dunn, D. D. 1996. Trends in Nutrient Inflows to the Gulf of Mexico from Streams Draining the Conterminous United States 1972-1993. U.S. Geological Survey, Water-Resources Investigations Report 96-4113. Prepared in cooperation with the U.S. Environmental Protection Agency, Gulf of Mexico Program, Nutrient Enrichment Issue Committee. Austin, Tex.: U.S. Geological Survey.

Federal Water Pollution Control Act (FWPCA) of 1948. 1948. Pub. L. No. 80-845, 62 Stat. 1155.

Federal Water Pollution Control Act (FWPCA) Amendments of 1961. 1961. Pub. L. No. 87-88, 75 Stat. 204, 208-09.

Federal Water Pollution Control Act (FWPCA) Amendments of 1972. 1972. Pub. L. No. 92-500, 86 Stat. 816.

Field, J. A., C. A. Johnson, and J. B. Rose. 2006. What is "emerging"? Environmental Science and Technology 40(23): 7105.

Florida Department of Environmental Protection (FDEP). 2003. Final Report: Integrating Atmospheric Mercury Deposition and Aquatic Cycling in the Florida Everglades: An Approach for Conducting a Total Maximum Daily Load Analysis for An Atmospherically Derived Pollutant. Tallahassee and Gainesville, Fla.

Fremling, C. R. 1964. Mayfly distribution indicates water quality on the Upper Mississippi River. Science 146: 1164-1166.

Fremling, C. R. 1989. *Hexagenia* mayflies: Biological monitors of water quality in the Upper Mississippi River. Journal of the Minnesota Academy of Sciences 55: 139-143.

Fremling, C. R. 2005. Immortal River: The Upper Mississippi in Ancient and Modern Times. Madison, Wisc.: The University of Wisconsin Press.

Fremling, C. R., and D. K. Johnson. 1990. Recurrence of *Hexagenia* mayflies demonstrates improved water quality in pool 2 and Lake Pepin, Upper Mississippi River. Pp. 243-248 in Mayflies and Stoneflies. I. C. Campbell (ed.). Proc. Int. Conf. Ephemeroptera, Vol. 5.

FTN Associates, Ltd., and Wenck Asssociates, Inc. 2005. Upper Mississippi River Fish Consumption Advisories: State Approaches to Issuing and Using Fish Consumption Advisories on the Upper Mississippi River. Prepared for the Upper Mississippi River Basin Association, St. Paul, Minn.

Galloway, J. N., and E. B. Cowling. 2002. Reactive nitrogen and the world: Two hundred years of change. Ambio 31: 64-71.

Galloway, J. N., J. D. Aber, J. W. Erisman, S. P. Seitzinger, R. W. Howarth, E. B. Cowling, and B. J. Cosby. 2003. The nitrogen cascade. BioScience 53: 341-356.

Garbarino, J. R., H. C. Hayes, D. A. Roth, R. C. Antweiler, T. I. Brinton, and H. E. Taylor. 1995. Heavy metals in the Mississippi River. Contaminants in the Mississippi River. R. Meade (ed.). U.S. Geological Survey Circular 1133. Denver, Colo.

Gebert, W. A., D. J. Graczyk, and W. R. Krug. 1987. Average annual runoff in the United States, 1951-1980. Madison, Wisc.: U.S. Geological Survey.

Gilbert, P. M., S. Seitzinger, C. A. Heil, J. M. Burkholder, M. W. Parrow, L. A. Codispoti, and V. Kelly. 2005. The role of eutrophication in the global proliferation of harmful algal blooms: New perspectives and new approaches. Oceanography 18(2).

Glanz, J. 1999. Sharp drop seen in erosion rates. Science 285: 1187-1189.

Goolsby, D. A. 2000. Mississippi Basin nitrogen flux believed to cause Gulf hypoxia. Eos, Transactions of the American Geophysical Union 81: 325-327.

Goolsby, D. A., and W. A. Battaglin. 2001. Long-term changes in concentrations and flux of nitrogen in the Mississippi River Basin, USA. Hydrologic Processes 15: 1209-1226.

Goolsby, D. A., and W. E. Pereira. 1995. Pesticides in the Mississippi River. Contaminants in the Mississippi River. R. Meade (ed.). U.S. Geological Survey Circular 1133. Denver, Colo.

Goolsby, D. A., W. A. Battaglin, G. B. Lawrence, R. S. Artz, B. T. Aulenbach, R. P. Hooper, D. R. Keeney, and G. J. Stensland. 1999. Flux and Sources of Nutrients in the Mississippi-Atchafalaya River Basin. Topic 3 Report for the Integrated Assessment on Hypoxia in the Gulf of Mexico. National Oceanic and Atmospheric Administration Coastal Ocean Program Decision Analysis Series No. 17. Silver Spring, Md.: National Oceanic and Atmospheric Administration Coastal Ocean Program. Available online at http://www.cop.noaa.gov/pubs/das/das17.pdf.

Greeley, W. B. 1925. The relation of geography to timber supply. Economic Geography 1: 1-14.

Hatch, L. K., A. P. Mallawatantri, D. Wheeler, A. Gleason, D. J. Mulla, J. A. Perry, K. W. Easter, P. Brezonik, R. Smith, and L. Gerlach. 2001. Land management at the major watershed-agroecoregion intersection. Journal of Soil and Water Conservation 56:44-51.

Headwaters Group Philanthropic Services. 2005. Traveling Upstream: Improving Water Quality of the Mississippi River. Prepared for the McKnight Foundation. St. Paul, Minn.

Hey, D. L., L. S. Urban, and J. A. Kostel. 2005a. Nutrient farming: The business of environmental management. Ecological Engineering 24: 279-287.

Hey, D. L., J. A. Kostel, A. P. Hurter, and R. H. Kadlec. 2005b. Comparative economics of nutrient management strategies: Traditional treatment and nutrient farming. Water Environment Research Foundation (WERF) Report #03-WSM-6CO, Alexandria, Va. Available online at http://www.wetlands-initiative.org/images/03WSM6COweb.pdf.

Houck, O. A. 1999. The Clean Water Act TMDL Program: Law, Policy, and Implementation, 2nd edition. Washington, D.C.: Environmental Law Institute.

Howarth, R. W., G. Billen, D. Swaney, A. Townsend, N. Jaworski, K. Lajtha, J. A. Downing, R. Elmgren, N. Caraco, T. Jordan, F. Berendse, J. Freney, V. Kudeyarov, P. Murdoch and Z. Zhao-Liang. 1996. Regional nitrogen budgets and riverine N & P fluxes for the drainages to the North Atlantic Ocean: Natural and human influences. Biogeochemistry 35: 75-79.

Hunt, C. B., and S. W. Trimble. 1998. Physiography of the United States. Pp. 864-884 in Reader's Encyclopedia of the American West. H. Lamar (ed.). New Haven, Conn.: Yale University Press.

Illinois Department of Natural Resources (ILDNR). 2006. Section 404 Program. Available online at http://dnr.state.il.us/wetlands/CH4B.HTM.

Ingram, R. 2006. The #1 Water Quality Issue for the Lower Mississippi River. Presentation to the NRC Committee on the Mississippi River and the Clean Water Act. Baton Rouge, La. May 11.

Interstate Council on Water Policy (ICWP). 2006. Interstate Water Solutions for the New Millennium. Washington, D.C.: Interstate Council on Water Policy.

Iowa Department of Natural Resources (IADNR). 2006. Section 401 Water Quality Certification. Available online at http://www.iowadnr.gov/water/section401/info.html.

Iowa Soybean Association (ISA). 2007. Environmental Programs. Available online at http://www.isafarmnet.com/ep/index.html.

Johnson, D. K. 2006. Personal communication. Metropolitan Council, Environmental Services, St. Paul, Minn.

Johnson, D. K., and P. W. Aasen. 1989. The metropolitan wastewater treatment plant and the Mississippi River: 50 years of improving water quality. Journal of the Minnesota Academy of Sciences 55: 134-138.

Justić, D., N. N. Rabalais, R. E. Turner, and W. J. Wiseman, Jr. 1993. Seasonal coupling between riverborne nutrients, net productivity, and hypoxia. Marine Pollution Bulletin 26: 184-189.

Justić, D., N. N. Rabalais, and R. E. Turner. 1996. Effects of climate change on hypoxia in coastal waters: A doubled CO_2 scenario for the northern Gulf of Mexico. Limnol. Oceanogr. 41: 992-1003.

Justić, D., N. N. Rabalais, and R. E. Turner. 2002. Modeling the impacts of decadal changes in riverine nutrient fluxes on coastal eutrophication near the Mississippi River Delta. Ecological Modelling 152: 33-46.

Justić, D., N. N. Rabalais, and R. E. Turner. 2003. Simulated responses of the Gulf of Mexico hypoxia to variations in climate and anthropogenic nutrient loading. Journal of Marine Systems 42: 115-126.

Kauffman, J., and W. Krueger. 1984. Livestock impacts on riparian ecosystems and streamside management implications: A review. Journal of Range Management 37: 430-438.

Keeney, D. R., and J. L. Hatfield. 2001. The nitrogen cycle, historical perspective, and current and potential future concerns. Pp. 3-16 in Nitrogen in the Environment: Sources, Problems, and Management. R. F. Follet and J. L. Hatfield (eds.). Cambridge, Mass.: Elsevier Science.

Kentucky Division of Water (KDOW). 2006. Division of Water: Kentucky Water Quality Certification Program. Available online at http://www.water.ky.gov/permitting/wqcert/General+Information.htm.

Kolpin, D. W., E. T. Furlong, M. T. Meyer, E. T. Thurman, S. D. Zaugg, L. B. Barber, and H. T. Buxton. 2002. Pharmaceuticals, hormones, and other organic wastewater contaminants in U.S. streams, 1999-2000: A national reconnaissance. Environmental Science and Technology 36: 1202-1211.

Kuechler, A. W. 1975. Potential Natural Vegetation of the Conterminous United States, 2nd Edition. Washington, D.C.: American Geographical Society.

Latacz-Loehmann, U., and C. van der Hamsvoort. 1997. Auctioning conservation contracts: A theoretical analysis and application. American Journal of Agricultural Economics 79: 407-418.

Leopold, L., M. Wolman, and J. Miller. 1964. Fluvial Processes in Geomorphology. San Francisco, Calif.: W. H. Freeman.

Lerczak, T. V., and R. E. Sparks. 1995. Fish populations in the Illinois River. Pp. 239-241 in National Biological Service National Status and Trends Report. Washington, D.C.: U.S. Government Printing Office. 530 pp.

Leue, A. 1886. The Forestal Relations of Ohio. First Annual Report of the Ohio State Forestry Bureau. Columbus, Ohio: Westbote Co., State Printers.

Lohrenz, S. E., G. L. Fahnenstiel, D. G. Redalje, G. A. Lang, X. Chen, and M. J. Dagg. 1997. Variations in primary production of northern Gulf of Mexico continental shelf waters linked to nutrient inputs from the Mississippi River. Marine Ecology Progress Series 155: 45-54.

Louisiana Department of Environmental Quality (LDEQ). 1999. Watershed Protection Programs—Mississippi River Basin. Available online at http://nonpoint.deq.state.la.us/99manplan/99mississippi.pdf.

Louisiana Department of Environmental Quality (LDEQ). 2006. Water Quality Certifications. Available online at http://www.deq.louisiana.gov/portal/tabid/2268/Default.aspx.

Lubowski, R., S. Bucholtz, R. Claassen, M. Roberts, J. Cooper, A Gueorguieva, and R. Johansson. 2006. Environmental Effects of Agricultural Land-Use Change: The Role of Economics and Policy. Economic Research Report Number 25. Washington, D.C.: Economic Research Service, U.S. Department of Agriculture.

Luzar, E. 1988. Natural resource management in agriculture: An institutional analysis of the 1985 Farm Bill. Journal of Economic Issues 22(2): 563-570.

Magnuson-Stevens Fishery Conservation and Management Act (MSA). 1976. Pub. L. No. 94-265, 90 Stat. 331.

Mamo, M., G. L. Malzer, D. J. Mulla, D. J. Huggins, and J. Strock. 2003. Spatial and temporal variation in economically optimum N rate for corn. Soil Science Society of America Journal 95: 958-964.

McPhee, J. 1999. The Control of Nature. New York: Farrar, Straus, and Giroux.

Meade, R. H. (ed.). 1995. Contaminants in the Mississippi River, 1987-1992. U.S. Geological Survey Circular 1133. U.S. Department of the Interior. Denver, Colo.: U.S. Geological Survey.

Meade, R. H., and R. Parker. 1985. Sediment in rivers of the United States. Pp. 49-60 in National Water Summary 1984. U.S. Geological Survey Water Supply Paper No. 2275.

Meade, R. H., T. Yuzyk, and T. Day. 1990. Movement and storage of sediment in rivers of the United States and Canada. Pp. 255-280 in The Geology of North America. W. H. Riggs (ed.). Geological Society of America.

Metropolitan St. Louis Sewer District. 2006. FAQ: CSO. Available online at http://www.stlmsd.com/EnvComply/ CSO/FAQ.1.html.

Milliman, J. D., and R. H. Meade. 1983. World-wide delivery of river sediment to the ocean. Journal of Geology 91: 1-21.

Minnesota Pollution Control Agency (MPCA). 2006. Clean Water Act Section 401 Water Quality Certifications. Available online at http://www.pca.state.mn.us/water/401.html#activities.

Minnesota Pollution Control Agency (MPCA). 2007. Lake Pepin Watershed TMDL Eutrophication and Turbidity Impairments Project Overview. Available online at http://www.pca.state.mn.us/water/tmdl.

Mississippi Department of Environmental Quality (MDEQ). 2006a. Wetlands Protection: Water Quality Certification Branch. Available online at http://deq.state.ms.us/MDEQ/nst/page/WQCB_Stream_ Wetland_Alteration03?OpenDocument.

Mississippi Department of Environmental Quality (MDEQ). 2006b. Mississippi 2006 Section 303(d) List of Impaired Water Bodies. Jackson, Miss.: Mississippi Department of Environmental Quality, Surface Water Division of the Office of Pollution Control. Available online at http://www.deq.state.ms.us/ MDEQ.nsf/pdf/TWB_2006_ 303d_List_draft_April_1_06/$File/MS_ 2006_Section_303d_List_ Draft_Version_April_1_2006.pdf?OpenElement.

Missouri Department of Natural Resources (MDNR). 1994. Basin 48: Mississippi River and Central Tributaries. Jefferson City, Mo.: State of Missouri Water Quality Basin Plan, Missouri Department of Natural Resources.

Missouri Department of Natural Resources (MDNR). 2006. 401 Water Quality Certification. Available online at http://www.dnr.mo.gov/env/wpp/401/index.html.

Mitsch, W. J., J. W. Day, Jr., J. W. Gilliam, P. M. Groffman, D. L. Hey, G. W. Randall, and N. Wang. 2001. Reducing nitrogen loading to the Gulf of Mexico from the Mississippi River basin: Strategies to counter a persistent ecological problem. BioScience 15: 373-388.

Moore, L. 2002. A Farm Level Analysis of the Economic Impacts of Federal Conservation Programs in the Midwest. M.S. Thesis. Lafayette, Ind.: Purdue University.

Mulla, D. J., A. S. Birr, N. Kitchen, and M. David. 2005. Evaluating the effectiveness of agricultural management practices at reducing nutrient losses to surface waters. Pp. 171-193 in Proceedings of the Gulf Hypoxia and Local Water Quality Concerns Workshop. Ames, Iowa: Iowa State University.

Nassar, R. 2006. Restoring America's Greatest River. Presentation to the NRC Committee on the Mississippi River and the Clean Water Act. Baton Rouge, La. May 11.

National Academy of Public Administration (NAPA). 2002. Understanding What States Need to Protect Water Quality. Washington, D.C.: National Academy of Public Administration.

National Advisory Council for Environmental Policy and Technology (NACEPT). 1998. Report of the Federal Advisory Committee on the Total Maximum Daily Load (TMDL) Program. EPA-100-R-98-006. Washington, D.C.: U.S. Environmental Protection Agency, National Advisory Council for Environmental Policy and Technology.

National Audubon Society. 1995. Investing in Wildlife; Multiple Benefits for Agriculture and the American People. Washington, D.C.: National Audubon Society.

National Research Council (NRC). 1999a. New Directions in Water Resources Planning for the U.S. Army Corps of Engineers. Washington, D.C.: National Academy Press.

National Research Council (NRC). 1999b. New Strategies for America's Watersheds. Washington, D.C.: National Academy Press.

National Research Council (NRC). 2000a. Clean Coastal Waters: Understanding and Reducing the Effects of Nutrient Pollution. Washington, D.C.: National Academy Press.

National Research Council (NRC). 2000b. Watershed Management for Potable Water Supply: Assessing the New York City Strategy. Washington, D.C.: National Academy Press.

National Research Council (NRC). 2001. Assessing the TMDL Approach to Water Quality Management. Washington, D.C.: National Academy Press.

National Research Council (NRC). 2002. The Missouri River Ecosystem: Exploring the Prospects for Recovery. Washington, D.C.: National Academy Press.

National Research Council (NRC). 2004a. Managing the Columbia River: Instream Flows, Water Withdrawals, and Salmon Survival. Washington, D.C.: The National Academies Press.

National Research Council (NRC). 2004b. River Basin and Coastal System Planning Within the U.S. Army Corps of Engineers. Washington, D.C.: The National Academies Press.

National Research Council (NRC). 2005. Water Resources Planning for the Upper Mississippi River and Illinois Waterway. Washington, D.C.: The National Academies Press.

National Research Council (NRC). 2007. Colorado River Basin Water Management: Evaluating and Adjusting to Hydroclimatic Variability. Washington, D.C.: The National Academies Press.

Natural Resources Conservation Service (NRCS). 2007. Conservation Effects Assessment Project (CEAP). Available online at http://www.nrcs.usda.gov/technical/nri/ceap/.

Nixon, R. M. 1970. Administration of Refuse Act Permit Program. Executive Order No. 11574. Federal Register 35: 19,627.

North Carolina Department of Environment and Natural Resources (NCDENR). 1998. Meuse River Basinwide Water Quality Plan. Available online at http://h2o.enr.state.nc.us/basinwide/Neuse/neuse_ wq_management_plan.htm.

Nowell, L. H., P. D. Capel, and P. D. Dileanis. 1999. Pesticides in stream sediment and aquatic biota—Distribution, trends, and governing factors. Pesticides in the Hydrologic System series, Vol. 4. 1040 pp. Boca Raton, Fla.: CRC Press. Available online at http://ca.water.usgs.gov/pnsp/abs/abp.html.

Pannell, R. P. 1999. Sediment Response to Large-Scale Environmental Change: The Upper Mississippi River, 1943-1996. Madison: University of Wisconsin. Unpublished master's thesis.

Parker, A. 2005. U.S. Environmental Protection Agency's National Nutrient Program. Available online at http://www.epa.gov/msbasin/taskforce/nutrient_workshop/ pdf/session-a.pdf.

Penland, S., R. Boyd, and J. R. Suter. 1988. The transgressive depositional systems of the Mississippi deltaic plain: A model for barrier shoreline and shelf sand development. Journal of Sedimentary Petrology 58(6):932-949.

Pennsylvania Department of Environmental Protection (PADEP). 2007a. Final Trading of Nutrient and Sediment Reduction Credits—Policy and Guidelines. Available online at http://www.depweb.state.pa.us/chesapeake/cwp/view.asp?a=3&Q=442886&chesapeakeNav=l29958.

Pennsylvania Department of Environmental Protection (PADEP). 2007b. PA Nutrient Trading. Available online at http://www.depweb.state.pa.us/chesapeake/lib/chesapeake/dec29_2006/finalpolicy_12-28.pdf.

Phillips, S. W., M. J. Focazio, and L. J. Bachman. 1999. Discharge, Nitrate Load, and Residence Time of Ground Water in the Chesapeake Bay Watershed. U.S. Geological Survey Fact Sheet 150-99. Available on-line at http://md.water.usgs.gov/publications/fs-150-99.

Pires, M. 2004. Watershed protection for a world city: The case of New York. Land Use Policy 21: 161-175.

Port of New Orleans. 2006. Facts about the Port of New Orleans. Available online at http://www.portno.com.

Prince, H. 1997. Wetlands of the American Midwest. Chicago, Ill.: University of Chicago Press.

Rabalais, N. N. 2002. Nitrogen in aquatic ecosystems. Ambio 31: 102-112.

Rabalais, N. N. 2005. Consequences of Mississippi River diversion for Louisiana Coastal Restoration. National Wetlands Newsletter (July-August): 21-24.

Rabalais, N. N., and R. E. Turner (eds.). 2001. Coastal Hypoxia: Consequences for Living Resources and Ecosystems. Coastal and Estuarine Studies 58. Washington, D.C.: American Geophysical Union.

Rabalais, N. N., and R. E. Turner. 2006. Oxygen depletion in the Gulf of Mexico adjacent to the Mississippi River. Pp. 225-245 in Past and Present Marine Water Column Anoxia. L. N. Neretin (ed.). NATO Science Series: IV-Earth and Environmental Sciences.

Rabalais, N. N., R. E. Turner, Q. Dortch, D. Justić, V. J. Bierman, Jr., and W. J. Wiseman, Jr. 2002a. Review. Nutrient-enhanced productivity in the northern Gulf of Mexico: Past, present and future. Hydrobiologia 475-476: 39-63.

Rabalais, N. N., R. E. Turner, and D. Scavia. 2002b. Beyond science into policy: Gulf of Mexico hypoxia and the Mississippi River. BioScience 52: 129-142.

Rabalais, N. N., R. E. Turner, B. K. Sen Gupta, E. Platon, and M. L. Parsons. 2007. Sediments tell the history of eutrophication and hypoxia in the northern Gulf of Mexico. Ecological Applications 17(5): S129-S143.

Randall, G. W., and D. J. Mulla. 2001. Nitrate nitrogen in surface waters as influenced by climatic conditions and agricultural practices. Journal of Environmental Quality 30: 337-344.

Randall, G. W., and J. A. Vetsch. 2005. Nitrate losses in subsurface drainage from a corn-soybean rotation as affected by fall application of nitrogen and nitrapyrin. Journal of Environmental Quality 34: 590-597.

Reagan, R. 1983. Exclusive Economic Zone of the United States of America. Proclamation No. 5030. Federal Register 48: 10,605.

Renaud, M. 1986. Hypoxia in Louisiana coastal waters during 1983: Implications for fisheries. Fishery Bulletin 84: 19-26.

Reuss, M. 1998. Designing the Bayous: The Control of Water in the Atchafalaya Basin, 1800-1995. Alexandria, Va.: U.S. Army Corps of Engineers, Office of History.

Ribaudo, M. 1986. Consideration of offsite impacts in targeting soil conservation. Land Economics 62: 402-411.

Ribaudo, M., R. Horan, and M. Smith. 1999. Economics of Water Quality Protection from Nonpoint Sources. Agricultural Economics Report No. 782. Washington, D.C.: U.S. Department of Agriculture, Economic Research Service.

Rostad, C. E., T. F. Rees, and S. R. Daniel. 1994. Colloid particle sizes in the Mississippi River and some of its tributaries, from Minneapolis to below New Orleans. Hydrological Processes 12(1): 25-41.

Royer, T. V., M. B. David, and L. E. Gentry. 2006. Timing of riverine export of nitrate and phosphorus from agricultural watersheds in Illinois: Implications for reducing nutrient loading to the Mississippi River. Environmental Science and Technology 40: 4126-4131.

Ruhl, J. B. 2000. Farms, their environmental harms, and environmental law. Ecology Law Quarterly 27(2): 263-349.

Sawyer, J., E. Nafziger, G. Randall, L. Bundy, G. Rehm, and B. Joern. 2006. Concepts and rationale for regional nitrogen rate guidelines for corn. PM 2015. Ames, Ia.: Iowa State University Extension.

Scavia, D., N. N. Rabalais, R. E. Turner, D. Justić, and W. J. Wiseman, Jr. 2003. Predicting the response of Gulf of Mexico hypoxia to variations in Mississippi River nitrogen load. Limnology and Oceanography 48:951-956.

Schertz, L., and O. Doering. 1999. The Making of the 1996 Farm Act. Ames, Iowa: Iowa State University Press.

Schumm, S. A., and B. R. Winkley (eds.). 1994. The Variability of Large Alluvial Rivers. New York: American Society of Civil Engineers Press.

Sekely, A. C., D. J. Mulla, and D. W. Bauer. 2002. Streambank slumping and its contribution to the phosphorus and suspended sediment loads of the Blue Earth River, Minnesota. Journal of Soil Water Conservation 57(5): 243-250.

Shabman, L., K. Stephenson, and W. Shobe. 2002. Trading programs for environmental management: Reflections on the air and water experiences. Environmental Practice 4 (September)3: 153-162.

Soballe, D. M. 1998. Successful water quality monitoring: The right combination of intent, measurements, interpretation, and a cooperating ecosystem. Lake and Reservoir Management 14(1): 10-20.

Soballe, D. M., and J. R. Fischer. 2004. Long Term Resource Monitoring Program Procedures: Water Quality Monitoring. U.S. Geological Survey, Upper Midwest Environmental Sciences Center, La Crosse, Wisc. Technical Report LTRMP 2004-T002-1 (Ref. 95-P002-5). 73 pp. + Appendixes A-J.

Soil and Water Conservation Society (SWCS). 2003. Comments on the Interim Rule for the Conservation Reserve Program, Soil and Water Conservation Society, July 7. Available online at http://www.swcs.org/en/ special_projects/farm_bill_conservation/.

Soil and Water Conservation Society (SWCS). 2006. Final Report from the Blue Ribbon Panel Conducting an External Review of the U.S. Department of Agriculture Conservation Effects Assessment Project. Ankeny, Iowa: Soil and Water Conservation Society.

Soil and Water Conservation Society (SWCS). 2007. Conservation Security Program (CSP) Program Assessment, Soil and Water Conservation Society (February). Available online at http://www.swcs.org/en/ special_projects/farm_bill_conservation/.

Soil and Water Conservation Society (SWCS) and Environmental Defense. 2007. Environmental Quality Incentives Program (EQIP) Program Assessment, Soil and Water Conservation Society, March. Available online at http://www.swcs.org/en/special_projects /farm_bill_conservation/.

Sparks, R. E., and A. J. Spink. 1998. Disturbance, succession and ecosystem processes in rivers and estuaries: Effects of extreme hydrologic events. Regulated Rivers 14(2): 155-177.

Spink, A., R. E. Sparks, M. van Oorschot, and J. A. Verhoeven. 1998. Nutrient dynamics of large river floodplains. Regulated Rivers 14(2): 203-216.

St. Paul Pioneer Press. 2006. Balance matters, But zoning is what protects the Mississippi.

Stavins, R. N. 2001. Experience with Market Based Environmental Policy Instruments. Discussion Paper 01-05. Washington, D.C.: Resources for the Future.

Stephenson, K., and L. Shabman. 2001. The trouble with implementing TMDLs. Regulation 24:1 (Spring): 28-32.

Stephenson, K., L. Shabman, and L. Geyer. 1999. Watershed-based effluent allowance trading: Identifying the statutory and regulatory barriers to implementation. Environmental Lawyer 5(June)3: 775-815.

Stoddard, A., J. B. Harcum, J. T. Simpson, J. Pagenkopf, and R. Bastien. 2002. Municipal Wastewater Treatment: Evaluating Improvements in Water Quality. Hoboken, N.J.: John Wiley and Sons, Inc.

Stoneham, G., V. Chaudrhi, A. Ha, and L. Strappazzon. 2003. Auctions for conservation contracts: An empirical examination of Victoria's Bush Tender Trail. Australian Journal of Agricultural and Resource Economics 47(4): 477-500.

Stow, C. A., S. S. Qian, and J. K. Craig. 2005. Declining threshold for hypoxia in the Gulf of Mexico. Environmental Science and Technology 39: 716-723.

Submerged Lands Act (SLA). 2006. 43 U.S.C. Section 1301-1356.

Tennessee Department of Environment and Conservation (TDEC). 2006. Environmental Permits Handbook: Aquatic Resource Alteration Permit. Available online at http://www.state.tn.us/environment/permits/arap/shtml.

Trimble, S. W. 1976. Unsteady state denudation. Science 191: 871.

Trimble, S. W. 1977. A sediment budget for the Coon Creek basin in the Driftless Area, Wisconsin, 1853-1975. American Journal of Science 283: 454-474.

Trimble, S. W. 1999. Decreased rates of alluvial sediment storage in the Coon Creek basin, Wisconsin, 1975-1993. Science 285: 1244-1246.

Turner, R. E., and N. N. Rabalais. 2003. Linking landscape and water quality in the Mississippi River basin for 200 years. BioScience 53: 563-572.

Turner, R. E., and N. N. Rabalais. 2004. Suspended sediment, C, N, P, and Si yields from the Mississippi River Basin. Hydrobiologia 511: 79-89.

Turner, R. E., N. Qureshi, N. N. Rabalais, Q. Dortch, D. Justić, R. F. Shaw, and J. Cope. 1998. Fluctuating silicate:nitrate ratios and coastal plankton food webs. Proceedings National Academy of Science 95: 13048-13051.

Turner, R. E., N. N. Rabalais, E. M. Swenson, M. Kasprzak, and T. Romaire. 2005. Summer hypoxia in the northern Gulf of Mexico and its prediction from 1978 to 1995. Marine Environmental Research 59: 65-77.

Turner, R. E., N. N. Rabalais, and D. Justić. 2006. Predicting summer hypoxia in the northern Gulf of Mexico: Riverine N, P and Si loading. Marine Pollution Bulletin 52: 139-148.

Turner, R. E., N. N. Rabalais, G. McIsaac, and R. W. Howarth. 2007 (accepted). Characterization of nutrient and organic matter loads from the Mississippi River to the Gulf of Mexico. Estuaries and Coasts.

United Nations Environmental Programme (UNEP). 2006. State of the Marine Environment. The Hague.

University of Maryland Center for Environmental Science (UMCES). 2007. Hypoxia Advisory Panel Draft Advisory. Cambridge, Md.

Upper Mississippi River Basin Association (UMRBA). 2004. Upper Mississippi River Water Quality: The States' Approaches to Clean Water Act Monitoring, Assessment, and Impairment Decisions. Water Quality Task Force. St. Paul, Minn.: Upper Mississippi River Basin Association.

Upper Mississippi River Basin Association (UMRBA). 2006. Organizational Options for Interstate Water Quality Management on the Upper Mississippi River. St. Paul, Minn.: Upper Mississippi River Basin Association.

Upper Mississippi River Conservation Committee (UMRCC). 2002. Upper Mississippi and Illinois River Floodplain Forests, Desired Future and Recommended Actions. Upper Mississippi River Conservation Committee, Rock Island, Ill. 35 pp.

Upper Mississippi River Water Suppliers Coalition (UMRWSC). 2006. Upper Mississippi River Water Suppliers Coalition. Available online at http://www.umrwsc.com.

U.S. Army Corps of Engineers (USACE). 1994. Economic Impacts of Recreation on the Upper Mississippi River System. St. Paul District, St. Paul, Minn.

U.S. Army Corps of Engineers (USACE). 2004. Integrated Feasibility Report and Programmatic Environmental Impact Statement for the UMR-IWW System Navigation Feasibility Study. Vicksburg, Miss.: Mississippi Valley Division Corps of Engineers.

U.S. Census Bureau. 2007. Population Change and Distribution, 1990-2000. Available online at http://www.census.gov/prod/2001pubs/c2kbr01-2.pdf.

U.S. Department of Agriculture (USDA). 2007a. USDA 2007 Farm Bill Proposals. Available online at www.usda.gov/farmbill.

U.S. Department of Agriculture (USDA). 2007b. Feed Grains Database: Yearbook Tables. USDA Economic Research Service. Available online at http://www.ers.usda.gov/Data/feedgrains/StandardReports/YBtable12.htm.

U.S. Environmental Protection Agency (USEPA). 1980. An Approach to Water Resources Evaluation on Non-Point Silvicultural Sources. EPA-600/8-80-012. Environmental Research Laboratory, Athens, Ga.

U.S. Environmental Protection Agency (USEPA). 1994. Combined Sewer Overflow Control Policy. Federal Register 59: 18,687.

U.S. Environmental Protection Agency (USEPA). 1998a. National Strategy for the Development of Regional Nutrient Criteria. Available online at http://www.epa.gov/waterscience/criteria/nutrient/nutsi.html.

U.S. Environmental Protection Agency (USEPA). 2001. Action Plan for Reducing, Mitigating, and Controlling Hypoxia in the Northern Gulf of Mexico. Mississippi River-Gulf of Mexico Watershed Nutrient Task Force, U.S. Environmental Protection Agency. Available online at http://www.epa.gov/msbasin/taskforce/actionplan.htm.

U.S. Environmental Protection Agency (USEPA). 2002. Summary Table for Nutrient Criteria Documents.

U.S. Environmental Protection Agency (USEPA). 2003a. Clean Watersheds Needs Survey 2000. Report to Congress, Facilities in Operation. Washington, D.C.: Office of Water, U.S. Environmental Protection Agency. Available online at http://www.epa.gov/owm/mtb/cwns/2000rtc/toc.htm

U.S. Environmental Protection Agency (USEPA). 2003b. Setting and Allocating the Chesapeake Bay Basin Nutrient and Sediment Loads: The Collaborative Process, Technical Tools and Innovative Approaches. EPA 903-R-03-007. Annapolis, Md: Region III Chesapeake Bay Program Office.

U.S. Environmental Protection Agency (USEPA). 2003c. Ambient Water Quality Criteria for Dissolved Oxygen, Water Clarity, and Chlorophyll A for the Chesapeake Bay and Its Tidal Estuaries. EPA 903-R-03-002. U.S. Environmental Protection Agency, Region III Chesapeake Bay Program Office.

U.S. Environmental Protection Agency (USEPA). 2003d. Technical Support Document for Identification of Chesapeake Bay Designated Uses and Attainability. EPA 903-R-03-004. U.S. Environmental Protection Agency, Region III Chesapeake Bay Program Office.

U.S. Environmental Protection Agency (USEPA). 2003e. Elements of a State Water Monitoring and Assessment Program. U.S. Environmental Protection Agency Report EPA 841-B-03-003. Available online at http://www.epa.gov/owow/monitoring/elements/ elements03 _14_03.pdf.

U.S. Environmental Protection Agency (USEPA). 2003f. Water Quality Trading Policy. Available online at http://www.epa.gov/owow/watershed/trading/finalpolicy.pdf.

U.S. Environmental Protection Agency (USEPA). 2004a. The Nonpoint Source Management Program: Pointer No. 4. EPA 841-F-96-004D. Available online at http://www.epa.gov/owow/nps/facts/point4.htm.

U.S. Environmental Protection Agency (USEPA). 2004b. Decision on Petition for Rulemaking to Publish Water Quality Standards for the Mississippi and Missouri Rivers Within Arkansas, Illinois, Iowa, Kansas, Kentucky, Missouri, Nebraska, and Tennessee. Letter from B. H. Grumbles, Office of Water, U.S. Environmental Protection Agency, Washington, D.C. to M. I. Lipeles, Interdisciplinary Environmental Clinic, Washington University in St. Louis, June 25.

U.S. Environmental Protection Agency (USEPA). 2006a. State NPDES Permit Program Authority. Available online at http://www.epa.gov/npdes/images/State_ NPDES_Prog_Auth.pdf.

U.S. Environmental Protection Agency (USEPA). 2006b. National Estuary Program: Programs in the Gulf of Mexico. Available online at http://www.epa.gov/owow/estuaries/programs/gom.htm.

U.S. Environmental Protection Agency (USEPA). 2006c. Watersheds. Available online at http://www.epa.gov/owow/watershed/.

U.S. Environmental Protection Agency (USEPA). 2006d. Letter from Benjamin H. Grumbles, Assistant Administrator to Doyle Childers, Director, Missouri Department of Natural Resources.

U.S. Environmental Protection Agency (USEPA). 2006e. Memorandum from Benjamin H. Grumbles, Assistant Administrator, to Regions 1-10, Establishing TMDL "Daily" Loads in Light of the Decision by the U.S. Court of Appeals for the D.C. Circuit in *Friends of the Earth, Inc. v. EPA*, et al., No. 05-5015 (April 25, 2006) and Implications for NPDES Permits, November 15.

U.S. Environmental Protection Agency (USEPA) 2006f. Water Quality Trading. Available online at http://www.epa.gov/owow/watershed/tradelinks.html.

U.S. Environmental Protection Agency (USEPA). 2006g. National Estuary Program. Available online at http://www.epa.gov/owow/estuaries/.

U.S. Environmental Protection Agency (USEPA). 2007a. Overview of EPA Authorities for Natural Resource Managers Developing Aquatic Invasive Species Rapid Response and Management Plans: CWA Section 404-Permits to Discharge Dredged or Fill Material. Available online at http://www.epa.gov/owow/invasive_species /invasives_management/cwa404.html.

U.S. Environmental Protection Agency (USEPA). 2007b. Monitoring and Assessing Water Quality: The National Water Quality Inventory Report to Congress 305(b) Report. Available online at http://www.epa.gov/305b/.

U.S. Environmental Protection Agency (USEPA). 2007c. Total Maximum Daily Loads. Available online at http://www.epa.gov/OWOW/tmdl/index.html.

U.S. Environmental Protection Agency (USEPA). 2007d. STORET Repository for Water Quality, Biological and Physical Data. U.S. Environmental Protection Agency, Office of Water, Washington, D.C. Available online at http://www.epa.gov/storet/index.html.

U.S. Environmental Protection Agency (USEPA). 2007e. About EPA Regions. Available online at http://www.epa.gov/epahome/locate2.htm.

U.S. Environmental Protection Agency (USEPA). 2007f. A Watershed Approach. Available online at http://www.epa.gov/owow/watershed/approach.html.

U.S. Environmental Protection Agency (USEPA) Inspector General. 2004. Effectiveness of Effluent Guidelines Program for Reducing Pollutant Discharges Uncertain. Washington, D.C.: Office of Inspector General, U.S. Environmental Protection Agency.

U.S. Environmental Protection Agency (USEPA), Region 4. 2002. Total Maximum Daily Load (TMDL) Development for Total Mercury in the Ochlockonee Watershed Including Listed Segments of the Ochlockonee River: Oquina Creek to Stateline/ State Route 37 Downstream Moultrie to Upstream CR222/ Bridge Creek to Big Creek. Available online at http://www.epa.gov/Region4/water/tmdl/georgia/ochlockonee/final_tmdls/OchlockoneeHgFinalTMDL.pdf.

U.S. Environmental Protection Agency (USEPA), Region 6. 2005. TMDLs for Mercury in Fish Tissue for Coastal Bays and Gulf Waters of Louisiana. U.S. EPA, Region 6, Dallas, Tex., and the Office of Environmental Assessment Louisiana Department of Environmental Quality. Contract 68-C-02-111. Prepared by Parsons. Dallas, Tex. Available online at http://www.epa.gov/waters/tmdldocs/6hgLATMDLs Report_05Jun28.pdf.

U.S. Fish and Wildlife Service (USFWS). 2007. Upper Mississippi River National Wildlife Refuge Comprehensive Conservation Plan. Available online at http://www.fws.gov/Midwest/planning/uppermiss/.

U.S. General Accounting Office (GAO). 2000a. Water Quality: Key EPA and State Decisions Limited by Inconsistent and Incomplete Data. GAO/RCED-00-54. Report to the Chairman, Subcommittee on Water Resources and Environment, Committee on Transportation and Infrastructure, House of Representatives, General Accounting Office.

U.S. General Accounting Office (GAO). 2000b. Environmental Protection: More Consistency Needed Among EPA Regions in Approach to Enforcement. GAO/SRCED-00-108. Washington, D.C.: U.S. General Accounting Office.

U.S. General Accounting Office (GAO). 2002. Water Quality: Inconsistent State Approaches Complicate Nation's Efforts to Identify its Most Polluted Waters. GAO-02-186. Report to Congress. Washington, D.C.: General Accounting Office.

U.S. General Accounting Office (GAO). 2003. Agricultural Conservation: USDA Needs to Better Ensure Protection of Highly Erodable Cropland and Wetlands. GAO-03-418. Washington, D.C.: U.S. General Accounting Office.

U.S. Geological Survey (USGS). 1999. Ecological Status and Trends of the Upper Mississippi River System 1998. K. Lubinski and C. Theiling (eds.). La Crosse, WI: U. S. Geological Survey, Upper Midwest Environmenal Sciences Center. Available online at http://www.umesc.usgs.gov/ products.html.

U.S. Geological Survey (USGS). 2006. National Stream Water Quality Network (NASQAN), Statistical Summaries of NASQAN Data. Available online at http://water.usgs.gov/nasqan/data/statsum.96.00.html.

U.S. Geological Survey (USGS). 2007. About the National Water Quality Assessment (NAWQA) Program. Available online at: http://water.usgs.gov/nawqa/about.html.

U.S. Government Accountability Office (GAO). 2006a. Securing Wastewater Facilities: Utilities Have Made Important Upgrades but Further Improvements to Key System Components May Be Limited by Costs and Other Constraints. GAO-06-390. Washington, D.C.: U.S. Government Accountabilty Office.

U.S. Government Accountability Office (GAO). 2006b. Clean Water: How States Allocate Revolving Loan Funds and Measure Their Benefits. Available online at http://www.gao.gov/cgi-bin/getrpt?GAO-06-579.

Vitousek, P. M., J. D. Aber, R. W. Howarth, G. E. Likens, P. A. Matson, D. W. Schindler, W. H. Schlesinger, and D. G. Tilman. 1997. Human alterations of the global nitrogen cycle: Sources and consequences. Ecological Applications 7(3): 737-750.

Vyn, T. 2007. Meeting the Ethanol Demand: Consequences and Compromises Associated with More Corn on Corn in Indiana. ID 336. Lafayette, Ind.: Purdue University Cooperative Extension Service. Available online at www.ces.purdue.edu/bioenergy.

Washington Post. 2004. Bay pollution progress overstated, July 18.

Water Pollution Control Act Amendments (WPCAA) of 1956. 1956. Pub. L. No. 84-660, 70 Stat. 498.

Water Quality Act (WQA) of 1965. 1965. Pub. L. No. 89-234, 79 Stat. 903.

Water Quality Act (WQA) of 1987. 1987. Pub. L. No. 100-4, 101 Stat. 7.

White, G. F. 1957. A perspective of river basin development. Law and Contemporary Problems 22(2): 157-184.

Wiebe, A. H. 1927. Biological survey of the Upper Mississippi River, with special reference to pollution. Bulletin of the U.S. Bureau of Fisheries 43(Part 2): 137-167.

Wiebe, K., and N. Gollehon. eds. 2006. Agricultural Resources and Environmental Indicators, 2006. Economic Information Bulletin No. EB-16. Washington, D.C.: Economic Research Service, U.S. Department of Agriculture.

Winger, P. V. 1986. Forested Wetlands of the Southeast: Review of Major Characteristics and Roles in Maintaining Water Quality. U.S. Fish and Wildlife Service Resource Publ. No. 163.

Wisconsin Department of Natural Resources (WDNR). 2006. Wetland Water Quality Certification. Available online at http://dnr.wi.gov/org/water/fhp/wetlands.shtml.

Wiseman, W. J., Jr., N. N. Rabalais, R. E. Turner, S. P. Dinnel, and A. MacNaughton. 1997. Seasonal and interannual variability within the Louisiana Coastal Current: Stratification and hypoxia. Journal of Marine Systems 12: 237-248.

Wortman, C. S., M. Helmers, A. Mallarino, C. Barden, D. Devlin, G. Pierzynski, J. Lory, R. Massey, J. Holz, C. Shapiro, and J. Kovar. 2005. Agricultural Phosphorus Management and Water Quality Protection in the Midwest. EPA Region VII. RP 187. Lincoln, Neb.: The Board of Regents of the University of Nebraska.

Wu, J., and W. Boggess. 1999. The optimal allocation of conservation funds. Journal of Environmental Economics and Management 38: 302-321.

Young, C., and P. Westcott. 2000. How decoupled is U.S. agricultural support for major crops? American Journal of Agricultural Economics 82(3): 762-767.

Young, T. F., and J. Karkoski. 2000. Green evolution: Are economic incentives the next step in non-point source pollution control? Water Policy 2: 151-173.

Zimmerman, R. J., and J. M. Nance. 2001. Effects of hypoxia on the shrimp fishery of Louisiana and Texas. Pp. 293-310 in Coastal Hypoxia: Consequences for Living Resources and Ecosystems. N. N. Rabalais and R. E. Turner (eds.). Coastal and Estuarine Studies 58. Washington, D.C.: American Geophysical Union.

Zucker, L. A., and L. C. Brown. 1998. Agriculture Drainage. Water Quality Impacts and Subsurface Drainage Studies in the Midwest. Ohio State University Extension Bulletin 871. Columbus, Ohio: Ohio State University.

Appendix A

Guest Speakers at Committee Meetings

Nonprofit Organizations and Trade Associations

Gretchen Bonfert, The McKnight Foundation, Minneapolis, Minnesota
Doug Daigle, Lower Mississippi Sub-basin Committee on Hypoxia, New Orleans, Louisiana
Jon Devine, Natural Resources Defense Council, Washington, D.C.
Jerry Enzler, National Mississippi River Museum and Aquarium, Dubuque, Iowa
Ted Heisel, Missouri Coalition for the Environment, St. Louis, Missouri
William Herz, The Fertilizer Institute, Washington, D.C.
Oliver Houck, Tulane University, New Orleans, Louisiana
Maxine Lipeles, Washington University, St. Louis, Missouri
Don Parrish, American Farm Bureau, Washington, D.C.
Richard Sparks, National Great Rivers Research and Education Center, Alton, Illinois
Holly Stoerker, Upper Mississippi River Basin Association, St. Paul, Minnesota
Nancy Stoner, Natural Resources Defense Council, Washington, D.C.
Roger Wolf, Iowa Soybean Association, Urbandale, Iowa

Federal Agencies

Richard Batiuk, U.S. Environmental Protection Agency, Annapolis, Maryland

David Bolgrien, U.S. Environmental Protection Agency, Duluth, Minnesota
William Franz, U.S. Environmental Protection Agency, Chicago, Illinois
Ron Nassar, U.S. Fish and Wildlife Service and Lower Mississippi River Conservation Committee, Vicksburg, Mississippi
Amy Parker, U.S. Environmental Protection Agency, Washington, D.C.
Diane Regas, U.S. Environmental Protection Agency, Washington, D.C.
Jeff Stoner, U.S. Geological Survey, Mounds View, Minnesota
Mike Sullivan, U.S. Natural Resources Conservation Service, Little Rock, Arkansas

State Agencies

Phil Bass, Mississippi Department of Environmental Quality, Jackson, Mississippi
Charles Correll, Iowa Department of Natural Resources, Des Moines, Iowa
Steven Heiskary, Minnesota Pollution Control Agency, St. Paul, Minnesota
Richard Ingram, Mississippi Department of Environmental Quality and Lower Mississippi River Conservation Committee, Vicksburg, Mississippi
Dean Lemke, Iowa Department of Agriculture and Land Stewardship, Des Moines, Iowa
Dugan Sabins, Louisiana Department of Environmental Quality, Baton Rouge, Louisiana
John Sullivan, Wisconsin Department of Natural Resources, La Crosse, Wisconsin
Tom VanArsdall, Kentucky Environmental and Public Protection Cabinet, Frankfort, Kentucky
Marcia Willhite, Illinois Environmental Protection Agency, East Springfield, Illinois
Jim Wise, Arkansas Department of Environmental Quality, Little Rock, Arkansas

Appendix B

Acronyms

ADEQ	Arkansas Department of Environmental Quality
ASIWPCA	Association of State and Interstate Water Pollution Control Administrators
BADT	Best available demonstrated technology
BAT	Best available technology
BMP	Best management practice
C2K	Chesapeake 2000
CAFO	Concentrated animal feeding operation
CALM	Consolidated Assessment and Listing Methodology
CASTNET	Clean Air Status and Trends Network
CEAP	Conservation Effects Assessment Project
CFR	Code of Federal Regulations
CFU	Colony-forming unit
CREP	Conservation Reserve Enhancement Program
CRP	Conservation Reserve Program
CSO	Combined sewer overflow
CSP	Conservation Security Program
CWA	Clean Water Act
CWRA	Clean Water Restoration Act
CWSRF	Clean Water State Revolving Fund
DDT	Dichlorodiphenyltrichloroethane
DOI	Department of the Interior

DRBC	Delaware River Basin Commission
EEZ	Exclusive Economic Zone
EMAP	Environmental Monitoring and Assessment Program
EMP	Environmental Management Program
EMTC	Environmental Management Technical Center
EPA	Environmental Protection Agency
EQIP	Environmental Quality Incentives Program
FCA	Fish Consumption Advisory
FDEP	Florida Department of Environmental Protection
FERC	Federal Energy Regulatory Commission
FOE	Friends of the Earth
FSA	Farm Service Agency
FWPCA	Federal Water Pollution Control Act
GAO	Government Accountability Office (formerly General Accounting Office)
IADNR	Iowa Department of Natural Resources
ICPRB	Interstate Commission on the Potomac River Basin
ICWP	Interstate Council on Water Policy
ILDNR	Illinois Department of Natural Resources
ISA	Iowa Soybean Association
KDOW	Kentucky Division of Water
LA	Load Allocation
LDEQ	Louisiana Department of Environmental Quality
LMRCC	Lower Mississippi River Conservation Commission
LTRMP	Long Term Resource Monitoring Program
MDEQ	Mississippi Department of Environmental Quality
MDNR	Missouri Department of Natural Resources
MPCA	Minnesota Pollution Control Agency
MSA	Magnuson-Stevens Act (of 1976)
NACEPT	National Advisory Council for Environmental Policy and Technology
NADP	National Atmospheric Deposition Program
NAPA	National Academy of Public Administration
NASQAN	National Stream Quality Accounting Network
NAWQA	National Water Quality Assessment

NCDENR	North Carolina Department of Environment and Natural Resources
NOAA	National Oceanic and Atmospheric Administration
NPDES	National Pollutant Discharge Elimination System
NRC	National Research Council
NRCS	Natural Resources Conservation Service
NRDC	Natural Resources Defense Council
NSPS	New Source Performance Standards
NTU	Nephelometric turbidity unit
ORSANCO	Ohio River Valley Water Sanitation Commission
PCB	Polychlorinated biphenyl
PCS	Permit compliance system
POTW	Publicly owned treatment work
PUD	Public Utility District
RBP	Rapid bioassessment protocol
SDWA	Safe Drinking Water Act
SLA	Submerged Lands Act
SRBC	Susquehanna River Basin Commission
SSO	Sanitary sewer overflow
STORET	Storage and Retrieval Environmental Data System
SWCS	Soil and Water Conservation Society
TDEC	Tennessee Department of Environment and Conservation
TMDL	Total Maximum Daily Load
TRI	Toxics Release Inventory
TSS	Total suspended solids
UMCES	University of Maryland Center for Environmental Sciences
UMRBA	Upper Mississippi River Basin Association
UMRCC	Upper Mississippi River Conservation Committee
UMR-IWW	Upper Mississippi River-Illinois Waterway
UMRSHNC	Upper Mississippi River Sub-basin Hypoxia Nutrient Committee
UMRWSC	Upper Mississippi River Water Suppliers Coalition
UNEP	United Nations Environment Programme
USACE	U.S. Army Corps of Engineers
USDA	U.S. Department of Agriculture
USFWS	U.S. Fish and Wildlife Service
USGS	U.S. Geological Survey

WDNR	Wisconsin Department of Natural Resources
WLA	Waste Load Allocation
WQA	Water Quality Act
WRDA	Water Resources Development Act
WSA	Wadable Streams Assessment
WSTB	Water Science and Technology Board

Appendix C

Biographical Information: Committee on the Mississippi River and the Clean Water Act

DAVID A. DZOMBAK (*chair*) is the Walter J. Blenko, Sr. Professor of Environmental Engineering and faculty director of the Steinbrenner Institute for Environmental Education and Research at Carnegie Mellon University. His research focuses on contaminant fate, transport, and treatment in water, soil, and sediment. Dr. Dzombak has published numerous articles in environmental engineering and science journals, book chapters, articles for the popular press, and two books. He serves on various national and regional committees, including the Environmental Protection Agency (EPA) Science Advisory Board (Environmental Engineering Committee), the EPA National Advisory Council for Environmental Policy and Technology (Environmental Technology Subcommittee), and the Southwestern Pennsylvania Regional Water Management Task Force. He is also an associate editor of the journal *Environmental Science & Technology*, a registered professional engineer in Pennsylvania, a diplomate of the American Academy of Environmental Engineers, and a fellow of the American Society of Civil Engineers. Dr. Dzombak holds a B.A. in mathematics from St. Vincent College, B.S. and M.S. degrees in civil engineering from Carnegie Mellon University, and a Ph.D. in civil engineering from Massachusetts Institute of Technology.

H. H. CHENG is professor emeritus and former head of the Department of Soil, Water, and Climate at the University of Minnesota. He was formerly a faculty member at Washington State University for 24 years (1965-1989), having served as associate dean of the Graduate School, interim chair of the Department of Agronomy and Soils, and chair of the Interdisciplinary Program in Environmental Science and Regional Planning. He served as professor and head of the department at the University of Minnesota from 1989 to 2002. He is a licensed professional soil scientist and a fellow of the Soil Science Society of America, the American Society of Agronomy, and the American Association for the Advancement of Science. He served as president of the Soil Science Society of America (1995-1996) and president of the American Society of Agronomy in 1999. His research interests include the chemistry, biochemistry, and analytical chemistry of soils; carbon and nitrogen cycles; transformation and transport of nitrogen, pesticides, and organic matter in the soil environment; crop residue management, nitrogen availability and use efficiency, and groundwater quality; residue and waste management; impact of climatic changes on carbon and nitrogen transformation dynamics; and precision agriculture and agricultural sustainability. He currently is a member of the National Research Council (NRC) Board on Agriculture and Natural Resources. Dr. Cheng received his Ph.D. from the University of Illinois in 1961, and the University of Minnesota conferred upon him a LL.D. (Hon.) degree in 2004.

ROBIN K. CRAIG is the Attorney's Title Insurance Fund Professor of Law at the Florida State University College of Law, Tallahassee. Prior to joining the law school in the fall of 2006, Dr. Craig was an associate professor of law and professor of law at the Indiana University School of Law, Indianapolis. She was a judicial clerk to Judge Robert E. Jones, U.S. District Court for the District of Oregon from 1996 to 1998, and also was a law clerk at the Oregon Department of Justice in the Natural Resources Section. Dr. Craig is active in the American Bar Association's Section on Environment, Energy, and Resources, Administrative Law Section, and Ocean Policy Working Group. She has authored two books, *The Clean Water Act and the Constitution* (Environmental Law Institute, 2004) and an environmental law textbook, *Environmental Law in Context* (West, 2005). Professor Craig also has written numerous law articles on the Clean Water Act, ocean and coastal law, and law and science, as well as the "Oceans and Estuaries" chapter of *Stumbling Toward Sustainability* (Environmental Law Institute, 2002), a review of the U.S. progress toward sustainable use of natural resources. Dr. Craig received her B.A. from Pomona College, her M.A. from Johns Hopkins University, her Ph.D. from the University of California, and her J.D. degree from Lewis & Clark School of Law.

OTTO C. DOERING III is a professor of agricultural economics at Purdue University. He is a public policy specialist and has served the U.S. Department of Agriculture (USDA) working on the 1977 and 1990 Farm Bills. In 1997 he was the principal advisor to USDA's Natural Resources Conservation Service for implementing the 1996 Farm Bill and served again in 2005 assisting in conservation program design and implementation. From 1985 to 1990 he was director of Indiana's State Utility Forecasting Group. In 1999 he was team leader for the economic analysis of the White House's National Hypoxia Assessment investigating the Dead Zone in the Gulf of Mexico. He serves on the EPA's Scientific Advisory Board Committee on Integrated Nitrogen. He is also president of the American Agricultural Economics Association. His publications include a book on the 1996 Farm Act and a book on the effects of climate change and variability on agricultural production systems. Other recent publications focus on economic linkages driving the response to nitrogen overenrichment, the rationale for U.S. agricultural policy, and the integration of biomass into existing energy systems. Dr. Doering received his B.A. degree from Cornell University, his M.S. degree from the London School of Economics, and his Ph.D. degree from Cornell University.

WILLIAM V. LUNEBURG, JR., is a professor of law at the University of Pittsburgh School of Law, where he teaches courses in environmental law, administrative law, civil procedure, and litigation with the federal government. He is the author of a variety of books, chapters, and articles dealing with issues of air and water pollution, among other topics. He developed his expertise in environmental law through his early work as an enforcement attorney with the U.S. Environmental Protection Agency. Professor Luneburg has consulted with or represented local environmental groups in litigation dealing with air and water pollution problems and forest preservation. He was a member of the EPA's Subcommittee on Ozone, Particulate Matter, and Regional Haze Implementation Programs that developed strategies to achieve the revised National Ambient Air Quality Standards for ozone and particulate matter adopted in 1997. For several years he was also a member of the Air Technical Advisory Committee of the Pennsylvania Department of Environmental Protection. Mr. Luneburg received his B.A. degree from Carleton College in political science and his J.D. degree from Harvard Law School.

G. TRACY MEHAN III is a principal with the Cadmus Group in Arlington, Va. He previously served as assistant administrator for water at the U.S. Environmental Protection Agency from 2001 to 2003; director of the Michigan Office of the Great Lakes and a member of the governor's cabinet from 1993 to 2001; and associate deputy administrator of EPA in 1992. Prior to that, he served as director of the Missouri Department of Natural Resources from 1989 to 1992. At EPA, he was known for using innovative approaches to protect drinking water and water resources. He was a leader on ambient water quality monitoring, the watershed approach, and a new strategy to deal with aging infrastructure. Mr. Mehan is the recipient of the 2004 Environment Award from the Association of Metropolitan Sewerage

Agencies (AMSA) and the 2003 Elizabeth Jester Fellows Environmental Partnership Award from the Association of State and Interstate Water Pollution Control Administrators (ASIWPCA). Mr. Mehan is a member of the Missouri Bar. Mr. Mehan received both his B.A. degree in history and his J.D. degree from St. Louis University.

JAMES B. PARK is the former chief of the Bureau of Water for the Illinois Environmental Protection Agency, where he spent a 30-year career implementing the Clean Water Act. Mr. Park was active in national policy discussions with the EPA and the ASIWPCA. He represented Illinois in the Ohio River Sanitation Commission (ORSANCO) and in the International Joint Commission for the Protection of the Great Lakes. He is a past chairman of the Board of Trustees for the Great Lakes Protection Fund, a trust established by the Great Lakes governors to fund research and other activities in the Great Lakes basin. He was also an active member of the Illinois River Coordinating Council, a group of state, federal, and citizen representatives responsible for developing action plans and policy initiatives for the Illinois River. Mr. Park received his B.S. and M.S. degrees in fluid mechanics from Southern Illinois University.

NANCY N. RABALAIS is the executive director of the Louisiana Universities Marine Consortium. Dr. Rabalais' research interests include the dynamics of hypoxic waters, interactions of large rivers with the coastal ocean, eutrophication, and benthic ecology. She is a fellow of the American Association for the Advancement of Science, an Aldo Leopold Leadership Program fellow, a past president of the Estuarine Research Federation, a national associate of the National Academies of Science, a vice chair of the Scientific Steering Committee of Land-Ocean Interactions in the Coastal Zone/International Geosphere-Biosphere Programme, and a past chair of the NRC Ocean Studies Board. She is currently on external advisory panels for the National Sea Grant Program and the National Science Foundation Environmental Biology and Education Directorate. She has authored 3 books, 26 book chapters, and more than 80 peer-reviewed publications. She received the 2002 Bostwick H. Ketchum Award for coastal research from the Woods Hole Oceanographic Institution and shares the Blasker award with R. E. Turner. She received her B.S and M.S. degrees in biology from Texas A&I University and her Ph.D. in zoology from the University of Texas.

JERALD L. SCHNOOR (NAE) is the Allen S. Henry Chair Professor in Engineering, professor in civil and environmental engineering; professor in occupational and environmental health, the College of Public Health; and co-director of the Center for Global and Regional Environmental Research at the University of Iowa. Dr. Schnoor is a member of the National Academy of Engineering and a registered professional engineer. His research interests are in mathematical modeling of water quality, phytoremediation, and global change. He has research projects on the design of environmental observatories, carbon sequestration to mitigate global warming, phytoremediation of hazardous wastes, and exposure risk assessment modeling. Dr. Schnoor is also editor-in-chief of the journal *Environmental Science and Technology*, co-editor of the John Wiley series of texts and monographs in *Environmental Science & Technology*, and a member of the EPA Science Advisory Board. He received his B.S. in chemical engineering from Iowa State University, his M.S. in environmental health engineering from the University of Texas, and his Ph.D. in civil engineering from the University of Texas.

DAVID M. SOBALLE is a research biologist with the U.S. Army Corps of Engineers. He has conducted more than 25 years of research in limnology, water quality, and river and reservoir ecology and has held research positions with state, federal, and academic institutions. Dr. Soballe has extensive experience working in interagency groups on water quality monitoring, data acquisition, and environmental management and restoration. He has expertise in the requirements and difficulties of monitoring a large floodplain river and in using monitoring data to guide management decisions. Dr. Soballe played a major role in redesigning and implementing the Long Term Resource Monitoring Program on the Upper Mississippi River, and for 12 years, 1991 through 2003, he led the water quality component of the Upper Mississippi River Environmental Management Program at the U.S. Geological Survey (USGS) Science Center in La Crosse, Wisconsin. He received his B.S. in biology in 1972 from the University of Notre Dame, his M.S. in biological sciences in 1978 from Michigan Technological University, and his Ph.D. in animal ecology (limnology) in 1981 from Iowa State University.

EDWARD L. THACKSTON is a professor of civil and environmental engineering, emeritus, at Vanderbilt University. His research interests include water quality in streams and reservoirs, mining and reaeration in streams, disposal of polluted dredged material, bacterial source tracking, and environmental law and policy. He taught water and waste treatment, water supply, wastewater collection, water quality management, and environmental law at Vanderbilt from 1965 through 2000, except for 1972-1973, when he was on leave to serve as staff assistant for environmental affairs to Tennessee Governor Winfield Dunn, for which he was named Tennessee Conservationist of the Year. From 1980 through 1999, he was chairman of the Department of Civil and Environmental Engineering at Vanderbilt. In 2001, he won the Landmark Paper Award from the Association of Environmental Engineering and Science Professors and was named Middle Tennessee Engineer of the Year by the Tennessee Society of Professional Engineers. Dr. Thackston received his B.E. in civil engineering from Vanderbilt University, his M.S. in sanitary engineering from the University of Illinois, and his Ph.D. in civil engineering from Vanderbilt University.

STANLEY W. TRIMBLE is a professor of geography at the University of California, Los Angeles. His research interests are conservation, hydrology, fluvial geomorphology, research techniques, and environmental history. He has been visiting professor at the Universities of Chicago, Vienna, Oxford, and London (University College) and was a research hydrologist with the U.S. Geological Survey from 1973 to 1984. Dr. Trimble has conducted field-based studies of geomorphology and sediment transport and storage issues in the Upper Mississippi River basin since the 1970s. His current field work includes investigations of these issues along the Mississippi River at Pool 8 near La Crosse, Wisconsin. Dr. Trimble has served on committees of the NRC, American Society of Civil Engineers, Association of American Geographers, and American Society for Testing and Materials. He is editor of the *Dekker Encyclopedia of Water Science* and joint editor-in-chief of *Catena*, an Elsevier journal of soils, hydrology, and landscape ecology. He is coauthor of the textbook *Environmental Hydrology*. In 2006, Dr. Trimble received the Mel Marcus Distinguished Career Award from the Geomorphology Specialty Group of the Association of American Geographers. Since 1978, he has owned and managed a 200-acre farm in Tennessee. Dr. Trimble received his B.S. in chemistry from the University of North Alabama and his M.S. and Ph.D. in geography from the University of Georgia.

ALAN H. VICORY, JR., is the executive director and chief engineer of the Ohio River Valley Water Sanitation Commission, an interstate compact water pollution control agency created to abate interstate water pollution in the Ohio Valley. After initial employment with Greeley & Hansen Engineers, where he was involved in water and wastewater studies and designs, Mr. Vicory joined the staff of ORSANCO, where he was appointed executive director and chief engineer in 1987. Mr. Vicory is a licensed professional engineer and is board certified in the area of water and wastewater by the American Academy of Environmental Engineers (AAEE). He is a past president of AEEE, a former chairman of the International Water Association's Watershed and River Basin Management Specialist Group, and past president of the Association of State and Interstate Water Pollution Control Administrators. Mr. Vicory received his B.S. in civil engineering from Virginia Military Institute in 1974.

STAFF

JEFFREY W. JACOBS is a senior program officer at the NRC's Water Science and Technology Board. Dr. Jacobs' research interests include policy and organizational arrangements for water resources management and the use of scientific information in water resources decision making. He has studied these issues extensively both in the United States and in mainland Southeast Asia. Prior to joining the NRC he served as a faculty member at the National University of Singapore and at Texas A&M University. Since joining the NRC in 1997, Dr. Jacobs has served as the study director for 20 NRC reports. He received his B.S. degree from Texas A&M University, his M.A. degree from the University of California, Riverside, and his Ph.D. degree from the University of Colorado.